STRUCTURE AND BONDING is issued at irregular intervals, according to the material received. With the acceptance for publication of a manuscript, copyright of all countries is vested exclusively in the publisher. Only papers not previously published elsewhere should be submitted. Likewise, the author guarantees against subsequent publication elsewhere. The text should be as clear and concise as possible, the manuscript written on one side of the paper only. Illustrations should be limited to those actually necessary.

Manuscripts will be accepted by the editors:

Professor Dr. *P. Hemmerich*   Universität Konstanz, Abteilung Biologie
                                 D-7750 Konstanz

Professor Dr. *C. K. Jørgensen*   51, Route de Frontenex,
                                    CH-1207 Genève

Professor *J. B. Neilands*   University of California, Biochemistry Department
                               Berkeley, California/USA

Sir *Ronald S. Nyholm*, FRS   Professor of Chemistry, University College
                                Gower Street
                                London WC 1/Great Britain

Professor Dr. *D. Reinen*   Institut für Anorganische Chemie der Universität
                              Marburg
                              D-3550 Marburg, Gutenbergstraße 18

Professor *R. J. P. Williams*   Wadham College, Inorganic Chemistry Laboratory
                                  Oxford/Great Britain

SPRINGER-VERLAG

D-6900 Heidelberg 1   D-1000 Berlin 33   SPRINGER VERLAG
P. O. Box 1780   Heidelberger Platz 3   NEW YORK INC.
Telephone (06221) 49101   Telephone (0311) 822001   175, Fifth Avenue
Telex 04-61 723   Telex 01-83319   New York, N. Y. 10010
                                                       Telephone 673-2660

STRUCTURE AND BONDING

Volume 8

Editors: P. Hemmerich, Konstanz
C. K. Jorgensen, Genève · J. B. Neilands, Berkeley
Sir Ronald S. Nyholm, London · D. Reinen, Marburg · R. J. P. Williams, Oxford

With 73 Figures

Springer-Verlag
Berlin Heidelberg GmbH 1970

ISBN 978-3-540-05257-9   ISBN 978-3-540-36403-0 (eBook)
DOI 10.1007/978-3-540-36403-0

The use of general descriptive names, trade marks, etc. in this publication, even if the former are not
especially identified, is not to be taken as a sign that such names, as understood by the Trade Marks and
Merchandise Marks Act, may accordingly be used freely by anyone.

This work is subject to copyright. All rights are reserved, whether the whole or part of the material is
concerned, specifically those of translation, reprinting, re-use of illustrations, broadcasting, reproduction
by photocopying machine or similar means, and storage in data banks. Under § 54 of the German
Copyright Law where copies are made for other than private use, a fee is payable to the publisher,
the amount of the fee to be determined by agreement with the publisher. © by Springer-Verlag Berlin
Heidelberg 1970 ·
Originally published by Springer-Verlag Berlin Heidelberg New York in 1970
Library of Congress Catalog Card Number 67-11280.

Contents

Contents

Iron Electronic Configurations in Proteins: Studies by Mössbauer Spectroscopy

A. J. Bearden and W. R. Dunham

Donner Laboratory, University of California, Berkeley, California 94720, USA

Table of Contents

I. Introduction

Resonance spectroscopy, which includes nuclear magnetic resonance, electron paramagnetic resonance, and nuclear gamma-ray resonance (Mössbauer spectroscopy) is a valuable adjunct to chemical and x-ray investigations of protein structure. The rapid utilization of high-frequency (220 MHz) proton magnetic resonance for the study of protein conformation in solution by *McDonald* and *Phillips* (*1, 2*) and by *Shulman et al.* (*3*), the role of electron paramagnetic resonance (EPR) in the study of paramagnetic active centers of proteins (for reviews see the works of *Beinert* and *Orme-Johnson* (*4, 5*) and by *Palmer et al.* (*6*)), and the exploitation of the "spin-label" technique for the study of proteins and membranes by *Hamilton* and *McConnell* (*7*), *McConnell et al.* (*8*), and *Hubbell* and *McConnell* (*9*) are familiar examples of the biological applications of resonance techniques. The use of Mössbauer spectroscopy in determining the electronic configurations at iron nuclei in iron-containing proteins is less general than other resonance methods, but has found usefulness in a number of iron-protein problems. The Mössbauer method measures small changes in the nuclear energy levels of a suitable Mössbauer nuclide; these energy changes are produced by the electronic surroundings. In a sense, the ^{57}Fe nuclide serves as a probe of extremely small dimension which does not perturb the electronic configuration as it transmits information about the surroundings. Unfortunately, there are no Mössbauer nuclides among the isotopes of C, N, O, P, S, or the biologically-interesting metals Mn, Mg, Cu, or Mo although the Mössbauer nuclides ^{127}I and ^{129}I may potentially be useful (*10*).

The early work in ^{57}Fe Mössbauer spectroscopy of biological molecules has been the subject of several reviews; for example, by *Bearden* and *Moss* (*11*), by *Phillips et al.* (*12*), and in a recent conference proceedings on this subject edited by *Debrunner, Tsibris,* and *Münck* (*13*). The two principal areas of application in biochemistry have been the study of hemoproteins and heme prosthetic groups, and the study of the iron-sulfur proteins. In both areas the goal has been to determine the electronic configurations at or near the active group by correlation of the Mössbauer spectroscopic results with other measurements, principally magnetic measurements by EPR or by magnetic susceptibility.

Before considering in review the present state of these researches and evaluating the effect of these researches on an understanding at the electronic level of these materials, it might be useful to present in an abbreviated form a discussion of the various parameters that are useful in describing Mössbauer spectroscopy. This will be done solely with the chemical and biochemical application in mind; more extensive reviews of Mössbauer spectroscopy are plentiful and delve into the physics of the

Mössbauer Effect and other areas of application more fully than will be done here (*14—18*). Despite the limitation that only biological materials containing iron are candidates for Mössbauer spectroscopy, there are a number of attractive features of this type of spectroscopy in comparison to other methods of spectroscopy, even other methods of resonance spectroscopy. First, there are no interfering signals from other atomic species; second, the Mössbauer nuclide does not perturb the protein from its normal configuration, objections of this sort have been raised concerning the use of spin-labels. The low natural abundance of the ^{57}Fe nuclide (2.19%) does suggest that Mössbauer spectra may be improved by either growing the host organism on an ^{57}Fe-enriched media (*19, 20*), pulse-enrichment by injection of an ^{57}Fe-containing metabolite (*21, 22*), or by undertaking chemical exchange either directly or by reconstitution of the protein with an ^{57}Fe enriched prosthetic group (*23, 24*). Mössbauer samples typically require 1—2 micromoles of ^{57}Fe in order to show "good contrast". It is also important to keep the concentration of atoms which absorb or scatter the 14.4 keV Mössbauer radiation low; sample volumes are typically less than 1 ml and ions such as Cl^- and Br^- are to be avoided because of the relatively high Compton scattering and photoelectric absorption present for these elements.

There is one other important factor which affects the design of a Mössbauer spectroscopic experiment; the sample must be in solid form. This is usually accomplished by using a frozen solution or lyophilizing the protein. Some attempts have been made to abtain Mössbauer spectra of iron-containing proteins in a sucrose solution of high viscosity (*25*) but these methods so far are not generally useful. The solid form requirement its imposed by the fact that small nuclear energy level changes can only be seen when the linewidth of the Mössbauer spectral lines are at a minimum and are determined by the properties of the nuclear transitions themselves and not affected by line-broadening due to recoil.

An important precaution for all low-temperature spectroscopy of biological materials has been pointed out by *Ehrenberg* (*26*). Protein conformations and electronic states are functions of temperature; therefore, it is important to ascertain whether the state under study of the material is the same, or even related to the state under physiological conditions. This precaution is in addition to the usual worries about harm being done to the protein by freezing and thawing or by lyophilization or buffer changes as a function of temperature. In several cases, most notably in the hydroxide derivatives of hemoglobin, magnetic states are a function of temperature or in the case of several forms of hemoglobin, are dependent on the degree of hydration of the protein (*24, 27, 28, 29*).

This article will limit its view to hemoproteins and heme prosthetic groups, the iron-sulfur proteins, and hemerythrin. Mössbauer spectroscopic studies have been incorporated into very thorough studies of iron-storage biomolecules; this field has already recently reviewed by *Spiro* and *Saltman* (*30*) and an additional effort here would be superfluous. Mössbauer spectroscopic studies of ferrichrome by *Wickman, Klein,* and *Shirley* are complete within themselves (*31—33*). Other iron-containing systems are under investigations; for many of these interpretation at this time is not as clear for hemoproteins and the iron-sulfur proteins: several of these studies will be mentioned briefly at the end.

II. Mössbauer Spectroscopy

A. Experimental Parameters

Mössbauer spectroscopy in biochemical application is normally arranged as a single-beam transmission spectroscopy: a source, an absorber containing a proportion of the Mössbauer nuclide (^{57}Fe), and a proportional counter or a scintillation counter as a detector of radiation. The source (^{57}Co) decays by K=electron capture to an excited state of ^{57}Fe; this state undergoes radiative decay to the low-lying 14.4 keV state which is used for the Mössbauer measurements. Sources are available commercially; source strengths in the range from 5 mCi to 60 mCi are normally employed. The 269 day halflife of ^{57}Co allows a single source to be used for a long series of experiments. The Mössbauer source must populate the *same* nuclide as contained in the resonant absorber. Some experiments have been carried out with the source in a protein rather than the absorber (*34*). Interpretation of these experiments rests on the details of the decay following K-electron capture and the establishment of a well-defined electronic state (*35, 36*).

Once again, it is important to state, perhaps more precisely, that the source and absorber must be in a solid form so that the gamma-ray energies are determined solely by the nuclear properties of the states and not by the recoil properties of individual atoms or molecules. Explicitly, the recoil energies are made vanishingly small by two factors: the incorporation of the source (or absorber) into a large coherent mass, the lack of lattice vibrations excited during the nuclear event. This latter condition is aided by lowering the temperature of the solid, 77 °K is a typical temperature for ^{57}Fe Mössbauer spectroscopy. Discussions of the basis of the Mössbauer effect are available at all levels of description; the more detailed works are by *Danon* (*37*) and *Kittel* (*38*).

Under these conditions the linewidth (ΔE) of the emitted (and absorbed) radiation is governed by the Heisenberg Uncertainty Principle; that is, $\Delta E = h/2 \pi \tau$ where h is Planck's radiation constant, and τ is the mean lifetime of the nuclear state. ($\tau = 1.4 \times 10^{-7}$ S for the first excited state of ^{57}Fe; this corresponds to an energy width of 10^{-9} eV.) In order for a spectroscopy to be useful some means must be introduced to change the energy of the beam; the absorbance is then recorded as a function of the variation in energy. In Mössbauer spectroscopy, a first-order relativistic energy shift is used; the addition of a small relative velocity between source and absorber causes an increase in the energy of the incident radiation. The equation, in first order, governing this energy shift is

$$\delta E = v \, E / c \, ,$$

where v is the relative velocity between source and absorber, E is the energy of the isomeric transition (14.36 keV for ^{57}Fe) and c is the velocity of light. Two conventions are defined: positive velocities correspond to motion of the source towards the absorber, and the convenient unit of velocity (mm/S) is used as an "energy" scale; 1 mm/S $= 4.8 \times 10^{-8}$ eV. The effects of the electronic configurations surrounding the Fe nuclide are generally from ten to several hundred times the natural linewidth as specified in the above velocity "energy" units. (The fullwidth at one-half maximum for an "unsplit" source absorber combination is 0.19 mm/S; the best spectrometers and well made sources approach 0.25 mm/S with the largest contribution to the increased width coming from inhomogeneity in the source.)

The interaction of the ^{57}Fe nuclide with the surroundings is through the magnetic and electric properties of the nuclide. The ^{57}Fe nuclide has a magnetic moment in both the $I = 1/2$ ground state (0.9024 nm) and the $I = 3/2$ excited state (-0.1547 nm). The charge densities of both states interact with the electronic charge density at the nuclear position (isomer shift); the $I = 3/2$ excited state has an electric quadrupole moment (0.3 barns) which can interact with electric field gradients produced by the surrounding electron charge densities. In keeping with terminology developed in the early days of optical spectroscopy, four basic interactions are possible for the ^{57}Fe nuclide. They are: the nuclear isomer (chemical) shift; the nuclear quadrupole interaction; and two magnetic interactions, the nuclear hyperfine interaction, and the nuclear Zeeman interaction.

The isomer shift illustrated in Fig. 1 is produced by small changes in the nuclear energy levels in both source and absorber brought about by the incorporation of the nuclide into an electron charge density. Only relative information about charge densities at the nuclear position is obtained; this being found from substitution of the protein absorber with another "standard" absorber such as Fe metal. For Fe(II) and Fe(III) configurations, isomer shift information is highly dependent on

5

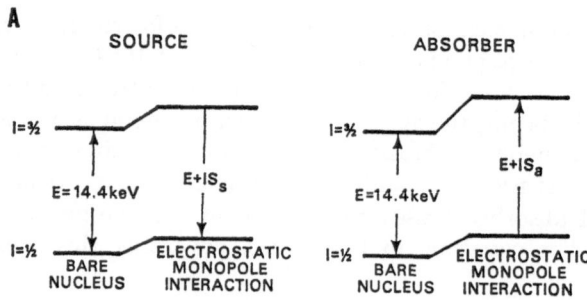

EFFECT OF ELECTRON CHARGE DENSITY

ON BARE NUCLEI

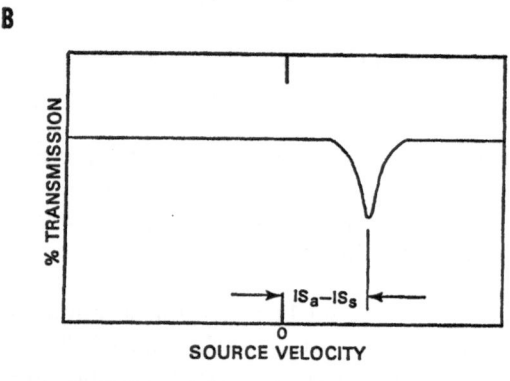

RESULTING MOSSBAUER SPECTRUM

Fig. 1. Mössbauer spectrum resulting from nuclear isomer shift

the details of the electronic configuration and the amount of electron delocalization or "back-donation" from surrounding ligands (14, 18, 39); in practice only high-spin Fe(II) is distinguished with certainty by its large (+ 1.5 to + 2.0 mm/S relative to the center of an Fe metal absorber value (40, 41). Theoretical treatment of isomer shifts for ^{57}Fe have been based on free-atom wave functions (42) and also introducing the effects of overlap and covalency (43).

The quadrupole interaction of the quadrupole moment of the I = 3/2 excited isomeric state of ^{57}Fe with an electric field gradient at the nuclear position produces a characteristic line pair in Mössbauer spectra as shown in Fig. 2. The energy difference is called the quadrupole splitting

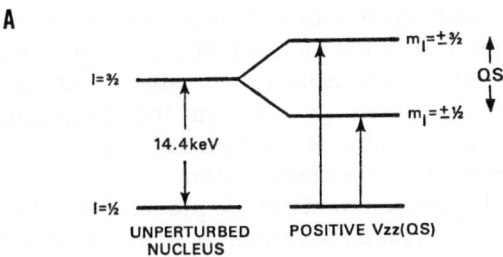

QUADRUPOLE SPLITTING OF NUCLEAR LEVELS

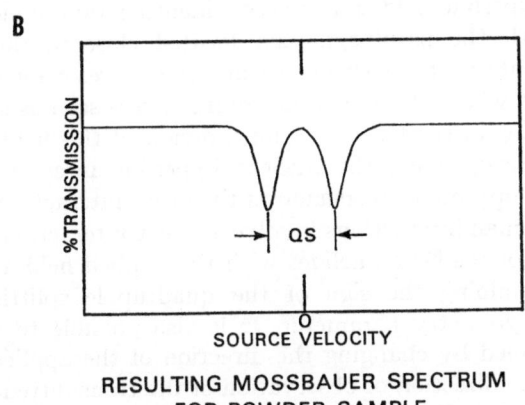

RESULTING MOSSBAUER SPECTRUM
FOR POWDER SAMPLE

Fig. 2. Mössbauer spectrum resulting from nonzero electric field gradient at the Mössbauer nuclear position for a powder sample

(44, 45). Two parameters describe the quadrupole interaction; there are the quadrupole splitting and the asymmetry parameter, η, which is defined as,

$$\eta = \frac{V_{xx} - V_{yy}}{V_{zz}}$$

where the double subscript denotes the second derivative of the potential. The sign of the quadrupole splitting and the value of the asymmetry parameter can be found by applying a large (30 to 50 kG) magnetic field to the sample (46, 47) and determining whether the $m_I = \pm 3/2$ states lie higher or lower in energy than the $m_I = \pm 1/2$ states. As in the case of

7

the isomer shift quadrupole interactions do not uniquely determine the electronic configuration of the Fe atom, only high-spin Fe(II) is uniquely determined, but unlike the isomer shift which contains only a single scalar parameter, the knowledge of the quadrupole splitting in both sign and magnitude and the value of the asymmetry parameter affords more opportunity to determine the configuration.

The nuclear hyperfine interaction is produced by the interaction of the nuclear moments in both states with an internal magnetic field produced at the nuclear position by one or more unpaired electrons (48). These internal fields may be quite large, 500 to 600 kOe for the five unpaired electrons in the $^6S_{5/2}$ state of the Fe(III) ion, even larger than the internal fields produced in a ferromagnet, 330 kOe in Fe metal. Observation of a nuclear hyperfine field is a positive identification of electron paramagnetism with the paramagnetism located close to the Mössbauer nuclide. In contrast to electron paramagnetic resonance studies in metalloproteins where the hyperfine interaction is seen as a small additional broadening of the EPR line (for a review of this field see: *Beinert* and *Orme-Johnson*, 1969), the nuclear hyperfine interaction in Mössbauer spectroscopy often overshadows the other interactions.

Nuclear Zeeman interactions involve a direct interaction of the magnetic moments of the ^{57}Fe nuclides with the applied field and are most useful in ascertaining the sign of the quadrupole splitting and the value of the asymmetry parameter. It is also possible to increase the information gained by changing the direction of the applied field from colinear with the direction of observation of the transmitted gamma ray to a direction perpendicular to the direction of observation; this change affects the relative transition probabilities between nuclear energy levels according to the rules for magnetic-dipole nuclear radiative transitions with the result that state assignments can be made based on the behavior of the absorbance (14, 15, 46, 32, 48).

In the foregoing discussion of the Mössbauer interactions two assumptions have been made. First, it has been implied that the interactions may be treated as first-order perturbations in the energy Hamiltonian for the system. This is a very good assumption; it is utilized in much the same manner as the "spin-Hamiltonian" approach in use in electron paramagnetic resonance spectroscopy (49, 50, 51). Secondly, all the interactions have been discussed as if they were time-independent. This is sometimes not the case thus complicating the interpretation of Mössbauer spectra. There are two important characteristic times that should be considered in discussions of Mössbauer spectroscopy, the mean lifetime of the nuclear excited state (1.4×10–S for ^{57}Fe) and the Larmor precession time of the nuclear magnetic moments which is inversely proportional to the internal magnetic field (32).

The quadrupole interaction and the isomer shift sense the electronic surroundings with a characteristic time which is the nuclear lifetime; thus measuring an average effect over this order of time. The fluctuations produced by molecular motions are generally more rapid than $10^7 - S^{-1}$, as molecular vibrations usually give rise to frequencies in the microwave region. What is important in any case is the magnitude of Fourier components having low frequencies; that is, in the order of $10^7 S^{-1}$ or lower. If there is appreciable amplitude to such a Fourier then one would expect to see additional quadrupole pairs or a line with a different value of the isomer shift with an absorbance proportional to the amplitude. Of course this absorbance would show a temperature dependence characteristic of the thermal population of the particular vibrational modes. Present analysis of Mössbauer spectra do not show any effects of this type, but there are a number of situations in solid-state physics where Mössbauer spectra might be expected to show the influence of "localized modes" (37).

The time dependent effects on the magnetic interactions are more complex. If the nuclear Larmor precession time is long compared to the nuclear radiative lifetime, then the nuclear magnetic energy states are not well defined. This means that, for example, it is impossible to see through magnetic transitions the effect of any field at the nucleus less than about 10 kG for ^{57}Fe. In addition, the internal field can be averaged to zero if in the nuclear hyperfine interaction the space quantization of the paramagnetic electron is disturbed by rapid enough spin-lattice or spin-spin relaxation processes. This effect can be minimized by working at low temperatures (4 °K) or by superimposing an external magnetic field and obtaining a large value of H/T. The decomposition of nuclear hyperfine interactions as the temperature is raised have been described for ferrichrome A by *Wickman, Klein,* and *Shirley* (31); a general discussion for many cases has also been given by *Wickman* (32).

In addition to the Mössbauer parameters which describe the perturbations of the nuclear energy levels and thereby give abundant information about the surrounding electronic configuration, there is an additional parameter, the Lamb-Mössbauer factor, which measures the probability that a "recoil free" event will occur either in the source or in the absorber (52). For ^{57}Fe Mössbauer spectroscopy it suffices to mention that the Lamb-Mössbauer factor approaches 0.8 at temperatures below 77 °K and is greater than 0.5 even at 300 °K. This permits the source to be kept at room temperature with cooling furnished only to the protein-absorber. Some attempts have been made to describe in more detail the Lamb-Mössbauer factor factor for ^{57}Fe (53, 54) but in general these correlations have not increased the amount of information that Mössbauer spectroscopy can give about a biochemical environment. In principle, such information might be useful, but the experimental arrangement required

9

is more than the usual transmission experiment (*55, 56*) or it is necessary that a very accurate measure of the amount of nonresonant radiation entering the detector be made. This last requirement is hard to realize for biomolecules as there is enhanced scattering of incident 122 keV as well as 14.4 keV radiation due to Compton processes from the high percentage of low-Z atoms present.

Finally, it should be mentioned that as single protein crystals are rarely used for Mössbauer spectroscopy in a biomolecular context, the general description is mainly directed at Mössbauer spectroscopy of a large and random collection of microcrystals such as found in a powder or in a frozen solution. Spatial averages taken under these experimental conditions usually average-out orientation-dependent Lamb-Mössbauer factors; the exception to this, the Goldanskii-Karyagin Effect (*57*) may arise from a number of possibilities, particulary from any effect which would give unequal Lamb-Mössbauer factors for different nuclear states. The origins and a thorough theoretical treatment of these effects have been given by Russian workers; at present, there are no biochemical conclusions which are dependent on a detailed knowledge of these Goldanskii-Karyagin Effects.

B. Theoretical Framework

Theoretical discussion of electronic configurations and correlation with Mössbauer spectra is usually undertaken within the framework of the "ligand field theory" approach by *Orgel* (*58*) and by *Ballhausen* (*59*), although "a crystal field" approach has been used in the analysis of some low-spin ferrihemoglobin compounds by *Harris-Lowe* (*60, 61*). In the "ligand field" method an energy level diagram has its rough qualitative features determined by field symmetry; for example, the five d-orbitals are split into two d_γ orbitals and three d_ε orbitals, the d_γ orbitals containing the $d_{x^2-y^2}$ and d_{z^2} orbitals and the d_ε orbitals containing the d_{xy}, d_{yz}, and d_{xz} orbitals. In octahedral symmetry the three d_ε orbitals lie lowest; the energy separation between the d_ε orbitals and the d_γ orbitals being designated as 10 Dq. In tetrahedral field symmetry the situation is reversed with the two d_γ orbitals lying lowest. For Fe(II) and Fe(III) there are six- and five-d-electrons respectively; these can be fitted into the orbitals in either a high spin, with maximum number of unpaired electrons, or a low spin, with maximum electron spin-pairing, scheme. The strength of the ligand field determines which of these two possibilities occurs; in the weak field or ionic case, the energy advantage for maximum exchange (spins parallel) produces the high-spin configuration; in a strong ligand field, the spin-paired low-spin configuration dominates. For ligand field symmetries other than octahedral or tetrahedral or for cases when appreciable rhombic or other distortions are present, the classification given above becomes less strict; for example, in the iron phthalocyanines, there is ample evidence for the existence of mid-spin states for ferrous iron (62, 63, 64). The energy splitting, Δ is larger for Fe(III) ions than Fe(II) ions in octahedral symmetry (58, 65); in tetrahedral symmetry, Δ is of the order of 4000 cm^{-1} for Fe(II) (66, 67) where the electron configuration is high spin. Because of the lower values for Δ in tetrahedral symmetry, Fe(II) low-spin configurations are unknown for this symmetry (58, 68). Detailed Mössbauer and magnetic susceptibility data and analysis have been made for a number of tetrahedrally-coordinated Fe(II) compounds (69, 70) and tetrahedrally-coordinated Fe(III) compounds (71).

In summary, high-spin Fe(II) is usually recognized in Mössbauer spectroscopy by the large values of the isomer shift (1—2 mm S) and the quadrupole splitting (2—4 mm S) and the pronounced temperature dependence of the quadrupole splitting produced by the unequal occupation of either d_γ or d_ε levels by the sixth d-electron. Mössbauer spectroscopy is a sensitive measure of distortions which split the d-orbitals, for example any rhombic distortion (70). High-spin Fe(III) in either octahedral or tetrahedral field symmetry is spherically symmetric in the free-ion limit ($^6S_{5/2}$). The small values of the quadrupole splitting and the isomer shift are then due to charges of surrounding ligands. These values of quadrupole splitting and isomer shift are not far different from those produced by low-spin Fe(III) compounds (21, 39) or low-spin Fe(II) compounds. The analysis of Mössbauer data for these states requires a consideration in detail of all the Mössbauer parameters; that is, the sign and magnitude of the quadrupole splitting, the asymmetry parameter, the presence or absence of nuclear hyperfine interactions, and the behavior of the spectra under applied magnetic field.

Iron atoms in states other than Fe(II) and Fe(III) are rare in biological material, but there is one case where Mössbauer evidence has pointed to an Fe(IV) electronic configuration. Horseradish peroxidase, when it forms peroxide derivatives (Compounds I and II of HRP), displays an isomer shift which is about equal to that obtained with Fe metal (23). A similar observation has also been found on an analogous compound, Japanese Radish Peroxidase (72). There is no evidence for Fe(I) or Fe(IV) states in any other hemoproteins, or in any of the iron-sulfur proteins.

It is important to realize that no one method of spectroscopy is clairvoyant. Electron paramagnetic resonance spectroscopy cannot sense low-spin Fe(II) as this state is diamagnetic, nor reliably the high-spin Fe(II) state because of rapid spin-lattice relaxation, large zero-field splittings or both; Mössbauer spectroscopy cannot distinguish Fe(III) spin states

11

A. J. Bearden and W. R. Dunham

without a detailed analysis over wide ranges of temperature and applied magnetic field; and magnetic susceptibility measurements give only the number of unpaired spins per molecule, not the location or spin species. The only worthwhile procedure is to gather the results from each of these methods done on the best available biochemical material and then to carefully make comparisons. It is this approach which has been followed in the study of two major classes of iron-proteins that will be discussed in the following sections.

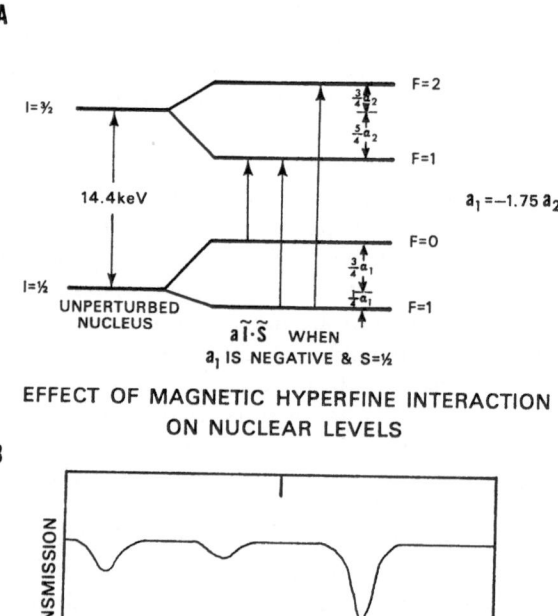

Fig. 3. Mössbauer spectrum resulting from a magnetic hyperfine interaction between electron paramagnetic moment and Mössbauer nuclide nuclear magnetic moments

III. Iron-Porphyrin Model Compounds

A. Fe(III) Hemin Compounds

Mössbauer spectroscopic investigations of hemin derivatives have been undertaken with both natural and enriched in ^{57}Fe compounds to provide model-system data for comparison with similar data on hemoproteins as well as to aid in the understanding of the electronic configurations of hemin compounds themselves (47, 73—78). These systems, formally Fe(III), have been studied with a variety of ring substitutions on the porphyrin structure; data on the proto-, meso-, and deutero- forms of hemin both in free acid form and as the dimethyl ester are the most common. The spectra show small isomer shifts (< 0.3 mm/S), but larger quadrupole splittings than normally would be expected for high-spin Fe(III) compounds (0.6 to 1.0 mm/S). There is small temperature dependence of both the isomer shift and the quadrupole splitting. The porphyrin-ring constraint distorts the electronic charge cloud of the Fe(III) configuration considerably from the spherical form of the free Fe(III) ion. As the fifth ligand is varied in a series of pentacoordinated hemins, the distortion of the symmetry increases following the series: fluoro- $<$ acetato- $<$ azido- $<$ chloro- $<$ bromo- $<$ (78). Experimentally, the sign of the electric field gradient is determined to be positive (47, 78, 81); this does not agree with the prediction of the molecular-orbital calculation by *Zerner et al.* (172), as is not surprising as "covalency effects" of the sort that are evident in this context are not apparent in a theoretical effort of this type. The large quadrupole splitting is in keeping with the x-ray data for hemin obtained by *Koenig* (79) in which the Fe ion is out-of-plane with respect to the porphyrin ring.

Normally Fe(III) high spin configurations present as dilute paramagnets show nuclear hyperfine interactions in zero applied magnetic field at temperatures above 4 °K (14); hemin chloride in frozen aqueous or frozen acetic acid solution does not. Instead there is an unusual behavior; the simple quadrupolar pair spectra obtains only at the lowest temperatures, generally below 30 °K. Above this temperature, one member of the pair shows line broadening although the area of this absorption remains equal to the area of the unmodified line. The explanation of this behavior is thought to rest in relaxation processes for the electronic spin which are available only at the higher temperatures. The EPR spectra of hemin compounds are all characterized by showing a $g_{\parallel} = 2$ and a $g_{\perp} = 6$. *Griffith* (80) has been able to account for this behavior by assuming a zero-field splitting for hemin of the order of 10 cm^{-1}; a value which is uncommonly large for Fe(III) ions, but which is due to square-planar arrangement of the porphyrin moiety. The existence of this large zero-field splitting was implied by the following considerations of the anomalous nuclear hyperfine interaction in the Mössbauer spectra of these materials and has now been measured directly with far infra-red absorption by *Feher, Richards,* and *Caughey* (*82, 83*) to be $2 D = 13.9$ cm^{-1}.

The zero-field splitting (2 D) separates the $S_z = \pm 1/2$ ground state of the system from the $S_z = \pm 3/2$ state; another energy-gap (4 D) separates this state from the higher $S_z = \pm 5/2$ state. At the lowest temperatures (4 °K) the $S_z = \pm 1/2$ state is populated exclusively and provides rapid relaxation of the electronic spin thereby averaging the nuclear hyperfine interaction to zero within the nuclear Larmor precession time. As the temperature is raised, the other S_z states become populated; the relaxation times are longer and may proceed by other processes; for example, spin-spin relaxation has been proposed by *Blume* (*84*). However, hyperfine fields are evident at the higher sample temperatures; this is what gives rise to the broadening of the line. This broadening identifies the broadened line as the predominantly $m_I = 3/2$ member of the pair (*78, 81*).

There have been two additional experiments which verified this basic picture of the nuclear hyperfine interaction in hemins. *Johnson* (*78*) increased the spin-lattice relaxation time by performing the Mössbauer experiment under field and temperature conditions which provide a large value of H/T. At 1.6 °K and in an applied field of 30 kG, a magnetic hyperfine interaction corresponding to that expected for high spin Fe(III) and for the g-values is measured experimentally. Recently, *Lang et al.* have found that a portion of hemin chloride dissolved in tetrahydrofuran at 1 mM concentration displays a hyperfine interaction at 4 °K in zero applied magnetic field. Their conclusion is that a portion of the hemin is present in a monomeric form in this solvent, a situation which is not apparent to any extent in water, acetic acid, chloroform, or dimethyl sulfoxide (*77*) at any concentrations used.

Chloroprotohemin in pyridine solution shows a different behavior; two quadrupole line pairs replace the single quadrupole line pair for chloroprotohemin in aqueous solution. The wide quadrupole splitting (1.8 mm/S) of the new line pair is more characteristic of Mössbauer spectra obtained for methemoglobin; these spectra will be discussed in the next section, but it suffices to point out here that the pyridine coordination produces an environment more nearly like the hexacoordinated environment of the iron in the hemoproteins (*78*).

B. Fe(II) Heme Compounds

Mössbauer data for 2,4 diacetyl deuteroporphyrin IX dimethyl ester and Fe(II) protoporphyrin IX obtained at 77 °K show isomer shifts similar to those for the Fe(III) hemin compounds in aqueous solution but larger quadrupole splittings (1.15 mm/S). In simple inorganic compounds which are ionic, reduction of Fe(III) usually strongly affects the isomer shift (40). However, covalent compounds such as ferrocyanide and ferricyanide show similar isomer shifts (40); this case has been treated theoretically by *Shulman* and *Sugano* (39) and the lack of isomer shift changes between these compounds has been interpreted in terms of delocalization of electrons through π-bonding. The conclusion for heme compounds is that similar delocalization obtains; this point will be discussed later as it applies to the oxygen-binding forms of the hemoproteins.

IV. Hemoproteins

A. Hemoglobin and Myoglobin

Mössbauer spectroscopic investigations on hemoglobin and myoglobin have been undertaken to support other magnetic investigations aimed at a complete electronic description of the heme site; hopefully in sufficient detail to permit the basis of oxygen-binding to be ascertained. For hemoglobin, the important feature; namely, the cooperative binding of oxygen must have as its root subtle configurational changes of the protein. Many techniques have been utilized to study this phenomena (85—94) including spin-label techniques (8). Mössbauer data has been acquired or many forms of myoglobin and hemoglobin (24, 95—108).

The most exhaustive series of studies, made with enriched compounds and with magnetic fields of up to 30 kG supplied by a superconducting solenoid, have been done by *Lang* and *Marshall* (101—104). Using the EPR g-values for low-spin forms of hemoglobin, a fit was possible to Mössbauer data taken at 4 °K and 1.2 °K as shown in Fig. 4. Positive assignment of a low-spin $S = 1/2$ configuration can be made as the result of these experiments. In contrast to the low-spin ferric environment, oxygenated hemoglobin displays a single quadrupole pair, unperturbed by any evidence of a nuclear hyperfine interaction, down to 4 °K. A small temperature dependence of the quadrupole splitting in oxyhemoglobin has suggested that the bound O_2 molecule may not be in the same orientation at all temperatures (103). The spin-state assignment on the basis of Mössbauer spectroscopy is not clear, but the data are compatible with an $S = 0$ assignment as determined by magnetic susceptibility measurements (103). The values of the isomer shift and the quadrupole splitting indicate a strong covalency between the ligand and the heme for both O_2-hemoglobin and NO-hemoglobin. There is no evidence that Mössbauer spectra of partially oxygenated hemoglobin shed any informa-

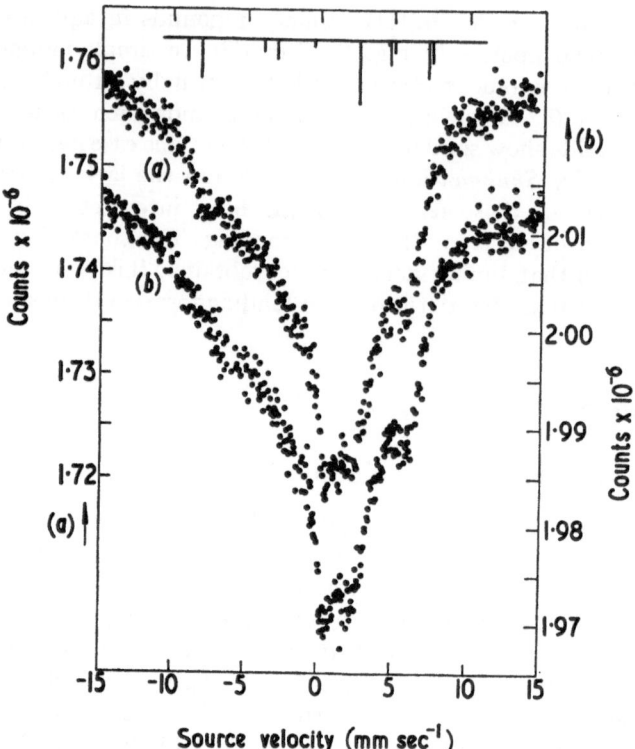

Fig. 4. Mössbauer absorption spectra of hemoglobin fluoride at a) 4 °K, background 40%, and b) 1.2 °K, background 40%.
Line spectrum calculated in the low temperature approximation, valid at both 4 °K and 1.2 °K. There are no free parameters; the lack of sharp lines in the observed spectra is attributed to spin relaxation (After *Lang* and *Marshall*, Ref. (*103*))

tion on the nature of cooperative effects producing the sigmoidal oxygenation curve for these compounds; perhaps the electronic environment at the Fe nuclear position is not sufficiently perturbed to sense the small changes associated with this phenomena.

High-spin ferric hemoglobin compounds, for example hemoglobin fluoride, display Mössbauer spectra characteristic of high-spin Fe(III) compounds but with large quadrupole splittings (1.8 mm/S) similar to hemin and a large nuclear hyperfine interaction which is particularly evident in spectra taken at the lowest temperatures (4 °K to 1.2 °K) (*103*). The data can be fit with a zero-field splitting (2 D) of about 14 cm^{-1} with g-values ($g_{\parallel} = 2$ and $g_{\perp} = 6$) which are not far from the parameters used for the analysis of high-spin hemin model compounds. Zero-field splittings of this size have been observed for high-spin forms of ferrihemoproteins by far infrared absorption techniques (*109, 110*).

Lyophilization of both ferrohemoglobin and ferrimyoglobin produces spin-state changes; the Fe(III) configuration in the dehydrated form of these proteins being low-spin in contrast to the high-spin form of the hydrated protein (*24, 111*). The evidence for this change in myoglobin is in the form of quadrupole data, a new pair of quadrupole lines appearing in the Mössbauer spectra with quadrupole splittings greater than 2 mm/S (*24*). This data supports the earlier assertion made on the basis of changes in the optical spectra of dehydrated methemoglobin (*112*) that spin-state changes do occur. *Williams* and coworkers have acquired Mössbauer data on a large series of hemoglobin compounds including the hydroxide which undergo spin-state changes as function of ligand or degree of hydration (*113*); it is clear that Mössbauer data is essential in following these phenomena.

An excellent review of the Mössbauer spectroscopic data on hemo-proteins has been published recently by *Lang* (*195*).

B. Cytochromes

Cytochromes from bacterial, yeast, and mammalian sources have been investigated by Mössbauer spectroscopy (*114—117*). Horseheart cyto-chrome *c* and the *c*-type cytochrome from *T. utilis* show spectra char-acteristic of low-spin Fe(III) in the oxidized form of the protein and low-spin Fe(II) for the reduced form of the protein. *Lang et al.* (*115*) have analyzed the Mössbauer data in terms of a low-spin Hamiltonian in some detail. *Cooke* and *Debrunner* (*116*) present quadrupole data on dehydrated forms of oxidized and reduced cytochrome *c*; the quadrupole splittings for hydrated and dehydrated forms of the reduced protein are quite similar in contrast to a difference of the oxidized form. No spin-state change is reported for either form of cytochrome *c*.

Mössbauer data for cytochrome *cc′* from *Chromatium* and *R. rubrum* and for cytochrome c_{552} from *Chromatium* and cytochrome c_2 from *R. rubrum* are reported in the work of *Moss et al.* (*117*) The oxidized cyro-chrome *cc′* data show highly-distorted high-spin electronic configurations similar to those in methemoglobin; the sign of the nuclear quadrupole coupling constant is reversed from that for methemoglobin indicating that the ligand binding is not similar. Cytochrome *cc′* in the reduced state is high-spin Fe(II) in contrast to the low-spin form found in the reduced cytochrome *c*.

Cytochromes *cc′* and methemoglobin hydroxide have been proposed to be proteins which also exist in a thermal mixture of S = 5/2 and S = 1/2 magnetic spin states (119). The fact that Mössbauer spectra taken over the temperature range from 4 °K to 270 °K consist of a single quadrupole pair has been stated as evidence that these states do not exist for times long compared to 10^{-8} S (101, 113, 24). It is important to point out that the large difference in the nuclear magnetic fields ($\sim$ 100 kOe as compared to 500 kOe) produced by $S_{z2} = \pm 1/2$ and $S_{z2} = \pm 5/2$ electronic states causes a large difference in the nuclear Larmor precession time necessary to observe the hyperfine field.

It has also been suggested that temperature-dependent quadrupole splittings as observed for methemoglobin and cytochrome cc' is evidence for thermally populated high- and low-spin states for these proteins. For measurements of quadrupole splittings, the salient nuclearly-imposed measurement time is the mean lifetime of the ^{57}Fe excited state, 1.4×10^{-7} S. Any change in the charge distribution surrounding the iron nucleus which is faster than 4×10^8 S^{-1} will be averaged; locally different charge environments which maintain the distribution for times long compared to 1.4×10^{-7} S will be sensed as distinct quadrupole environments. Vibrational oscillations for the usual ligands bound to iron in metalloproteins are at microwave frequencies; that is, within 10^{-7} S there will be many oscillations so that the Mössbauer measurements sense an average of the surrounding charge distributions.

C. Horseradish Peroxidase and Japanese Radish Peroxidase

Horseradish peroxidase is a particularly interesting material for Mössbauer examination as it has been reported that there are two oxidizing equivalents above the Fe(III) state (119, 120). A green Compound I is formed when the enzyme if reacted with hydrogen peroxide or with alkyl hydroperoxides (121). The red Compound II is formed more slowly from Compound I and retains one oxidizing equivalent above Fe(III) (120, 122). Magnetic susceptibility measurements by Theorell and Ehrenberg (123) were compatible with the assignment of Fe(V) for Compound I and Fe(IV) for Compound II. The Mössbauer studies of Moss et al. (23) show that HRP-Compound I is in a Fe(IV) state, but that Compound II does not display Mössbauer spectra characteristic of a Fe(V) state. Similar studies of Japanese Radish peroxidase by Maeda (72) show the same result although the work of Morita and Mason (124) show differences in the magnetic properties of the two enzymes and the intermediate compounds.

D. Cytochrome C Peroxidase

Perhaps the best example of a well-defined nuclear hyperfine splitting in a hemoprotein is the Mössbauer spectra of cytochrome c peroxidase-fluoride obtained by *Lang (125)*. When either the meso- or the proto-forms are examined at 4 °K with a small (100 G) magnetic field used to align the electronic magnetic moments, the spectra shown in Fig. 5 are the result. The assignment of a high-spin Fe(III) state similar to that in methemoglobin fluoride with a zero-field splitting of about 7 cm^{-1} is unequivocal. When cytochrome c peroxidase is examined at pH values nearer neutrality there is evidence for a mixture of high- and low-spin states.

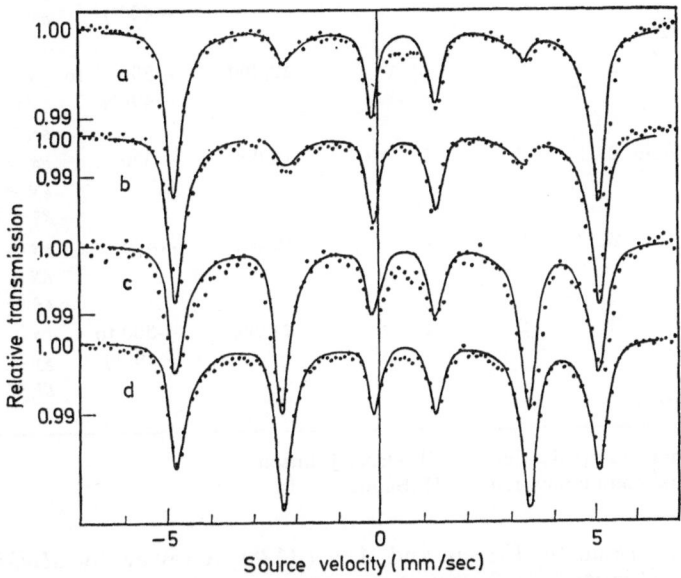

Fig. 5. Mössbauer spectra of cytochrome c peroxidase fluoride; (a) — proto, (b) — meso. A magnetic field of 100 G is applied parallel to the direction of observation of the gamma-ray beam; (c) — proto, (d) — meso. A magnetic field of 500 G is applied perpendicular to the direction of observation of gamma-ray beam. (After Lang, Ref. (125)).

V. Iron-Sulfur Proteins

The second class of iron-containing proteins which have been well-studied by Mössbauer spectroscopy, and by other resonance techniques, are the iron-sulfur proteins. These molecules are also known by the name, ferredoxins. Iron-sulfur proteins in several varieties serve as electron-transport agents for processes in plants, bacteria, and mammals. Perhaps the most-studied physiological process involving the iron-sulfur proteins is the study of their role in photosynthesis. This subject has been exten-sively reviewed by *Arnon (126, 135)*, *Hind* and *Olson (127)*, *Hall* and

Table 1

	Fe	S=	Electrons Transferred	MW	E_0' (mV)	EPR g-values
Azotobacter Fe—S Protein I	2	2	1	21,000	−300 to −400[a])	$gx = 1.93$ $gy = 1.95$ $gz = 2.02$
Azotobacter Fe—S Protein II	2	2	1	24,000	−300 to −400[a])	$gx = 1.90$ $gy = 1.95$ $gz = 2.05$
Parsley Ferredoxin	2	2	1	12,000	−300 to −400[a])	$gx = 1.89$ $gy = 1.96$ $gz = 2.05$
Adrenodoxin	2	2	1	12,000	−370	$gx = 1.93$ $gy = 1.94$ $gz = 2.02$
Spinach Ferredoxin	2	2	1	12,000	−420	$gx = 1.89$[b]) $gy = 1.96$[b]) $gz = 2.05$[b])
C. Pasteurianum Paramagnetic Protein	2	2	1	24,000	−300 to −400[a])	$gx = 1.93$ $gy = 1.95$ $gz = 2.00$

[a]) Personal communication, W. H. Orme-Johnson
[b]) Personal communication, R. H. Sands

Evans (*194*) and by *Vernon* and *Avron* (*128*). A review by *Malkin* and *Rabinowitz* (*129*) summarizes the properties of the iron-sulfur proteins, and in particular discusses the work on ferredoxins linked to nonphotosynthetic processes; this involvement of ferredoxin was implied in the earliest researches by *Mortenson*, *Valentine*, and *Carnahan* (*130*) and by *Tagawa* and *Arnon* (*131*).

The progress of research on the structure of the active site of the iron-sulfur proteins can be summarized by dividing the discussion into three parts: a discussion of the basically biochemical studies aimed at determining the chemical properties of the iron-sulfur moiety, the unsuccessful attempts so far to determine the structure by means of x-ray crystallography, and the application of resonance techniques as an alternate method of determining the configuration of the iron-sulfur complex. The iron and inorganic sulfide content, the reactivity of the complex, and a refutation of the claim by *Bayer*, *Parr*, and *Kazmaier* (*132*) that additional sulfide is not necessary to reconstitute ferredoxin from apoferredoxin are presented in a series of researches by *Buchanan*, *Lovenberg*, and *Rabino-*

witz (133) and by *Malkin* and *Rabinowitz (134)*. These matters are covered in the review mentioned above *(129)*. Unfortunately, to date, although small crystals can be obtained of the iron-sulfur proteins *(126)*, crystals of a size sufficient for x-ray use have not been available. The methods of isomorphic replacement which have been essential for x-ray studies have also not been workable with these materials *(136)*.

Iron-sulfur proteins can be classified on the basis of iron and sulfide content into "plant-type" iron-sulfur proteins, and "bacterial-type" iron-sulfur proteins. Plant-type iron-sulfur proteins contain just two Fe and two inorganic S atoms per protein molecule; the name refers to the material first isolated from chloroplasts. The bacterial-type iron-sulfur proteins always contain more than two Fe (and S^{--}) atoms per protein molecule; in general there are eight Fe and eight S^{--} atoms per protein molecule.

All iron-sulfur proteins, whether of the plant-type or the bacterial-type have three characteristics in common: all contain the acid-labile sulfide in equimolar ratio to iron; all show reduction potentials in the range from -240 to -420 mV (E_0', $pH = 7.0$); and when these proteins are chemically-reduced (typically with dithionite), they display an uncommon EPR signal, known as the "$g = 1.94$" signal. The oxidized forms of the proteins are not paramagnetic *(159)*.

A. Plant-Type Iron-Sulfur Proteins

Table 1 lists some of the properties of the plant-type iron sulfur-proteins for which extensive study by EPR and Mössbauer spectroscopy has been reported. The physical properties summarized show that the plant-type iron sulfur proteins have molecular weights in the range from 12,000 to 24,000 and have EPR g-values (g_x, g_y, g_z) all of the "$g = 1.94$" type shown in Fig. 6 but with minor variations reflecting axial or nonaxial symmetry of the paramagnetic center. The amino-acid sequences of four plant-type iron-sulfur-proteins are known: alfalfa *(136)*, *I.. glauca (137)*, *Scenedesmus (138)*, and spinach *(139)*. Each protein has about 97 residues, all in a single peptide chain; these are shown in Table 2.

The plant-type iron-sulfur proteins can accept a single electron from substrate *(140)*; this single electron is accounted for quantitatively by double-integration of the EPR signal in the reduced form of the protein *(140, 141)*. In addition to providing a structure for the active paramagnetic center of the protein, resonance spectroscopy may provide a picture of the electronic configuration; this would answer questions as how the conformation of the protein prevents a second electron from being accepted from substrate and a description of the magnetic properties of the protein in both the oxidized and reduced forms of the protein.

Table 2

	1	2	3	4	5	6	7	8	9	10	11	12	13	14
1. L. Glauca	----	Ala-	Phe-	Lys-	Val-	Lys-	(Leu)-	Leu-	Thr-	Pro-	Asp-	Gly-	(Pro)-	Lys-
2. Spinach	Ala		Tyr			Thr		Val			Thr		Asn	Val
3. Alfalfa	Ala	Ser	Tyr					Val			Glu		Thr	Gln
4. Scenedesmus	Ala	Thr	Tyr			Thr		Lys			Ser		Asp	Gln

↓ (arrow at 18)

	15	16	17	18	19	20	21	22	23	24	25	26	27	28
1. L. Glauca	Glu-	Phe-	Glu-	Cys-	Pro-	Asp-	Asp-	Val-	Tyr-	Ile-	Leu-	Asp-	Gln-	Ala-
2. Spinach		Gln												Ala
3. Alfalfa														His
4. Scenedesmus	Thr	Ile							Thr					Ala

↓ (arrow at 39)

	29	30	31	32	33	34	35	36	37	38	39	40	41	42
1. L. Glauca	Glu-	Glu-	Leu-	Gly-	Ile-	(Asp)-	Leu-	Pro-	Tyr-	Ser-	Cys-	Arg-	Ala-	Gly-
2. Spinach			(Glu)		(Ile)									
3. Alfalfa			Glu				Val							
4. Scenedesmus			Ala		Leu									

↓↓ (arrows at 44 and 47)

	43	44	45	46	47	48	49	50	51	52	53	54	55	56
1. L. Glauca	Ser-	Cys-	Ser-	Ser-	Cys-	Ala-	Gly-	Lys-	Leu-	Val-	Glu-	Gly-	Asp-	Leu-
2. Spinach									Lys	Thr		Ser		
3. Alfalfa									Val	Ala	Ala	Glu	Val	
4. Scenedesmus	Ala								Val	Glu	Ala	Thr	Val	

	57	58	59	60	61	62	63	64	65	66	67	68	69	70
1. L. Glauca	Asp-	Gln-	Ser-	Asp-	Gln-	Ser-	Phe-	Leu-	Asp-	Asp-	Glu-	Gln-	Ile-	Glu-
2. Spinach	Asn		Asp								Asp		Asp	
3. Alfalfa			Ser		Gly						Asp			
4. Scenedesmus											Ser		Met	Asp

↓ (arrow at 77)

	71	72	73	74	75	76	77	78	79	80	81	82	83	84
1. L. Glauca	Glu-	Gly-	Trp-	Val-	Leu-	Thr-	Cys-	Ala-	Ala-	Tyr-	Pro-	Arg-	Ser-	Asp-
2. Spinach													Val	
3. Alfalfa								Val		Ala	Lys			
4. Scenedesmus	Gly		Phe			Val				Thr				

	85	86	87	88	89	90	91	92	93	94	95	96	97
1. L. Glauca	Val-	Val-	Ile-	Glu-	Thr-	His-	Lys-	Glu-	Glu-	Glu-	Leu-	Thr-	(Ala)
2. Spinach		Thr											
3. Alfalfa		Thr											
4. Scenedesmus	Cys	Thr		Ala						Asp		Phe	----

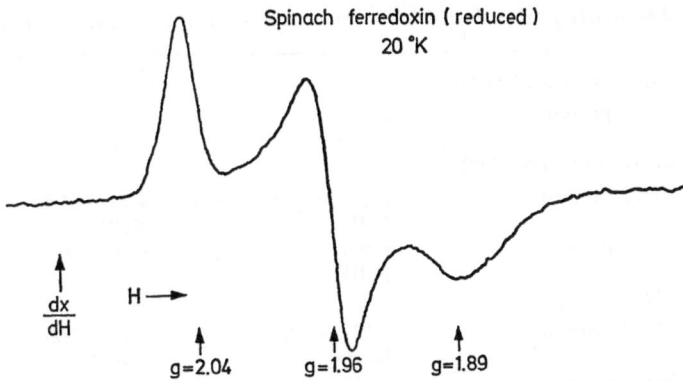

Fig. 6. Electron paramagnetic resonance signal showing the "$g = 1.94$" characteristic of the dithionite-reduced spinach ferredoxin, a plant-type iron-sulfur protein. Spectrum taken at 20 °K

The $g = 1.94$ EPR signal exhibited in the reduced state of the ferredoxins was the basis for models of the active site of these proteins. The identification of this EPR signal with an iron complex has been described in a review by *Beinert* and *Palmer* (*142*). The complexity of the iron ligand field which is necessary to produce a $g = 1.94$ signal was demonstrated by *Beinert et al.* (*143*), who proposed a model compound for this signal. This model compound was pentacyanonitrosylferrate (I). The properties of this model compound were later related and expanded by *Van Voorst* and *Hemmerich* (*144*).

Meanwhile, Blumberg and *Peisach* (*145*) showed that the interaction between a low-spin ferrous atom and an adjacent free radical can give rise to a $g = 1.94$ EPR signal. *Brintzinger, Palmer*, and *Sands* (*146*) proposed the first two-iron model for the active center of a plant-type ferredoxin. Their model, which consisted of two spin-coupled, low-spin ferric atoms in the oxidized protein and one low-spin ferric and one low-spin ferrous atom in the reduced protein, explained much of the chemical data on the proteins. Later, they (*Brintzinger, Palmer*, and *Sands,* (*147*)) presented EPR data for a compound, bis-hexamethylbenzene, Fe(I), which demonstrated all the properties fo the $g = 1.94$ signal observed in the ferredoxins.

The above model was criticized by *Gibson et al.* (*148*), and *Thornley et al.* (*150*), who reported that the tetrahedral symmetry of the BPS model could not give the crystal-field splitting required for spin pairing in the iron atoms. They, instead, proposed a model with two high-spin ferric atoms in the oxidized protein which were exchange coupled to render this state diamagnetic. In the reduced state, their model consists of a ferric (S = 5/2) state exchange coupled to a ferrous (S = 2) state to

Table 3. *Previously-proposed models for the active center of the plant-type ferredoxins*

A. *Blumberg* and *Peisach* (*145*)

Reduced protein	Fe^{II}	—	Free
	$(S = 0)$		Radical

B. *Brintzinger et al.* (*146, 147*)

Oxidized protein	↑ $(S_1 = 1/2)$	←→	↓ $(S_2 = 1/2)$
	Fe^{III}		Fe^{III}
Reduced protein	↑ $(S_1 = 1/2)$	←→	o$(S_2 = 0)$
	Fe^{III}		Fe^{II}

C. *Gibson et al.* (*148*)

Oxidized protein	↑ $(S_1 = 5/2)$	—	↓ $(S = 5/2)$
	Fe^{+++}		Fe^{+++}
Reduced protein	↑ $(S_1 = 5/2)$	—	↓ $(S = 2)$
	Fe^{+++}		Fe^{++}

D. *Johnson et al.* (*149*)

Fe^{+++} ↑	Fe^{+++} ↓	e^-	Fe^{++} o	Fe^{+++} ↑
$\overline{\quad(S=O)\quad}$		→	$\overline{\quad(S=1/2)\quad}$	
low-spin	low-spin		low-spin	low-spin
Fe^{++} o	Fe^{++} o		Fe^+ ↑	Fe^{++} o
low-spin	low-spin	e^-	low-spin	low-spin
$\overline{\quad(S=O)\quad}$		→	$\overline{\quad(S=1/2)\quad}$	
Oxidized Protein			Reduced Protein	

give a resultant spin to the complex as a whole of $S = 1/2$. Thus, the reduced state was ferrimagnetic, and they attributed the high-temperature disappearance of the EPR signal to two-phonon Orbach processes (*151*). The $g = 1.94$ signal was explained by assuming a tretrahedral ligand field about the ferrous atom with spin-orbit coupling constant of $75\ cm^{-1}$. This model explained all the properties of the $g = 1.94$ EPR signal; also, it has the advantage of being quite plausible in view of the known sulfur ligands around the iron atoms.

Several Mössbauer spectroscopic papers have dealt with members of the plant-type ferredoxins. In these papers, the Mössbauer spectra for a particular protein were interpreted to yield information such as the oxidation state and spin state of the iron atoms in the protein, and in some cases this information was extended to validate a proposed model for the active site. However, problems with denatured protein material or incorrect interpretation of the Mössbauer data have prevented any of these models from being accepted as valid.

Bearden and *Moss* (*12*) and *Moss et al.* (*152*) presented the Mössbauer spectra of spinach ferredoxin in its oxidized and reduced states. These spectra showed the two iron atoms in the oxidized protein in identical electronic environments. Upon protein reduction, one of the iron atoms exhibited a spectrum characteristic of a high-spin ferrous ion. The

Mössbauer spectra of the reduced proteins in the above study are not consistent with subsequent data for these proteins (*153, 164, 165*). It is now believed (personal communications, *W. H. Orme-Johnson* and *Graham Palmer*) that 1. the samples in these experiments were impure, and 2. the buffers used in these experiments were not strong enough to maintain a buffer *p*H level during the dithionite reductions. Therefore, the Mössbauer spectra of reduced spinach ferredoxin in the above experiment resulted from a mixture of oxodized protein iron and iron from denatured protein material.

Johnson et al. (*154, 155*) interpreted the spectra on spinach ferredoxin (similar to those of *Moss et al.* (*152*)) as consistent with the following interpretation: a) the oxidized protein contains two low-spin ferrous ions, and b) the reduced protein contains one low-spin ferrous ion and one high-spin ferrous ion. *Cooke et al.* (*156*) interpreted their data (similar to the data contained in the present work) on putidaredoxin in the following manner: a) the electronic environments of both iron atoms are identical in the oxidized protein, with the diamagnetism of this material resulting from spin pairing between the iron atoms, and b) in the reduced state, a single electron is shared equally by both iron atoms and gives rise to the internal magnetic field observed in the Mössbauer spectra. *Novikov et al.* (*157*) have published the results of a Mössbauer spectroscopic study on an iron-sulfur protein from *Azotobacter*. Both the data and the conclusions are similar to those made by *Moss et al.* (*152*) on spinach ferredoxin. Recently, *Johnson et al.* (*155*) and *Johnson et al.* (*149*) have published Mössbauer studies on the ferredoxins from *Euglena* and spinach. They now report their data as being most favorable to two models for the active site of these proteins.

Fig. 7 shows the Mössbauer spectra obtained by *Dunham et al.* (*153*) of the oxidized state of all the plant-type ferredoxins. The isomer shift and quadrupole splittings for these spectra are listed below:

Table 4. *Mössbauer parameters for the oxidized proteins*

	IS/Pt[a] (mm/S)	QS (mm/S)	η
Spinach Fd.	-0.08 ± 0.01	0.65 ± 0.01	0.05 ± 0.2
Parsley Fd.	-0.07 ± 0.01	0.66 ± 0.01	0.05 ± 0.2
Adrenodoxin	-0.08 ± 0.01	0.61 ± 0.01	0.05 ± 0.2
Putidaredoxin	-0.08 ± 0.01	0.61 ± 0.01	0.5 ± 0.2
Clos. Paramag. Protein	-0.07 ± 0.01	0.62 ± 0.01	0.5 ± 0.2
Azoto. Fe —S Protein I	-0.04 ± 0.01	0.73 ± 0.01	0.5 ± 0.2
Azoto. Fe —S Protein II	0.06 ± 0.01	0.71 ± 0.01	0.5 ± 0.1

[a]) Isomer shifts quoted here are given relative to a gamma ray source consisting of [57]Co diffused into a platinum matrix.

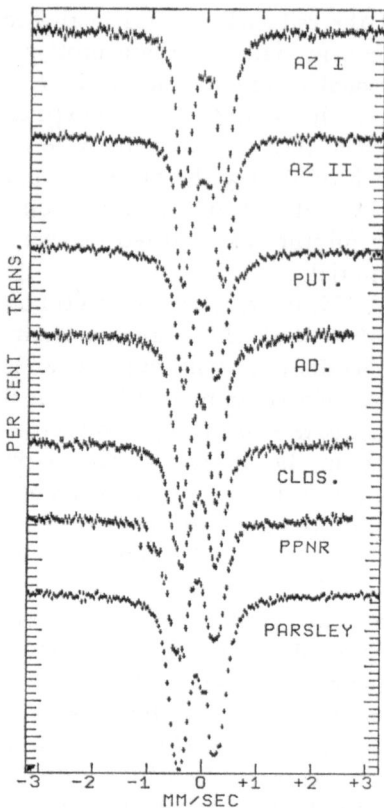

Fig. 7. Mössbauer spectra of oxidized plant-type iron-sulfur proteins in zero applied magnetic field. Abbreviations: AZI = *Azotobacter* Fe-S protein I, 4.6 °K; AZII = *Azotobacter* Fe-S protein II, 4.2 °K; Put. = Putidaredoxin, 4.2 °K; Ad. = Pig Adrenodoxin, 4.2 °K; Clos. = Clostridial paramagnetic protein, 4.2 °K; PPNR = Spinach ferredoxin, 4.5 °K; Parsley = Parsley Ferredoxin, 4.2 °K. Velocity scale is relative to iron in platinum

The parameters, IS and QS, shown in Table 5 are measured at 4.2 °K with zero applied field. The value of η and the sign of QS are derived by matching computed spectra to the Mössbauer data for the oxidized proteins taken at 4.2 °K in 46 kilogauss applied magnetic field (Fig. 8). The above parameters do not exhibit any measurable temperature dependence over the temperature range from 4.2 °K to 77 °K.

Thus, the best fit to the oxidized protein data is a single quadrupole pair with an isomer shift of $-.08$ mm/S and an observed splitting of 0.65 mm/S.

Table 5. *Mössbauer parameters for the high temperature reduced PPNR spectra*

	IS/Pt (mm/S)	QS (mm/S)	η
Iron Site # 1	−0.10	+0.64	0.5
Iron Site # 2	+0.19	−2.63	0 to 0.5

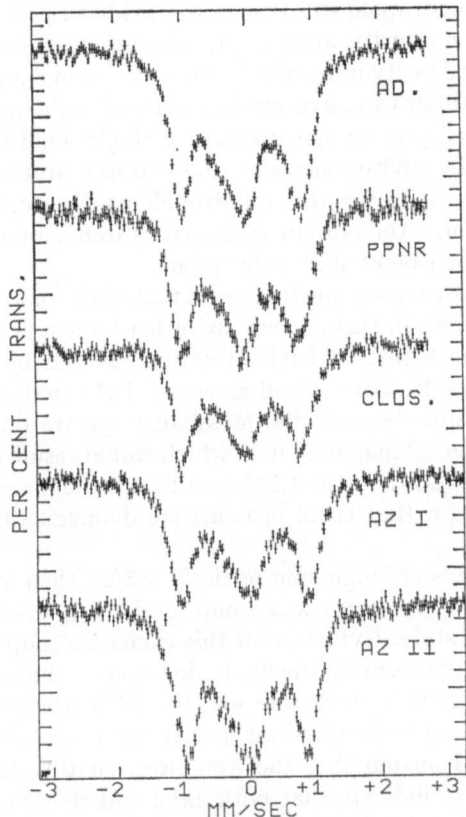

Fig. 8. Mössbauer spectra of oxidized plant-type iron-sulfur proteins in high applied magnetic field. Abbreviations: Ad. = Pig Adrenodoxin, 4.2 °K, 46 kG; PPNR = Spinach Ferredoxin, 4.5 °K, 50 kG; Clos. = Clostridial Paramagnetic Protein, 4.2 °K, 46 kG; AZI = *Azotobacter* Fe-S Protein I, 4.6 °K, 46 kG; AZII = *Azotobacter* Fe-S Protein II, 4.2 °K, 46 kG. Applied magnetic field is parallel to gamma-ray direction

The most probable electron configurations for iron atoms in a ligand field formed by amino-acid side chains and sulfur are d^5 and d^6. The crystal field splitting (Δ) required to pair spins in iron compounds is

A. J. Bearden and W. R. Dunham

greater than 15.000 cm^{-1} (59). Ligand field theory calculations (65) indicate that even in octahedral coordination, strong field ligands are required to cause spin pairing of iron atoms. The only side chain capable of supplying this strong field ligand is histidine. Since there is only one histidine common to the plant-type ferredoxins (Table 2) and since sulfur is shown to be a ligand in the iron complex, low-spin iron configurations are doubtful for these proteins.

The small quadrupole splitting in the oxidized protein spectra imply that the electron density around the iron atoms is nearly spherical. A spherical charge density indicates that the iron is an S-state ion, although low-spin ferric atoms can have small quadrupole splittings (158). In addition, the oxidized protein spectra show a single quadrupole pair, which indicates that the environments for the two iron atoms are nearly identical. The isomer shift for this quadrupole pair is most consistent with that of ferric iron, although low-spin ferrous iron cannot be ruled out as possibility by the isomer shift value alone.

Thus, the most reasonable interpretation of the oxidized protein data is that the iron sites in this protein are either high-spin ferric or low-spin ferrous, with the high-spin ferric situation favored by the ligand field arguments set forth above. Combinations of electrons for the two iron site are not possible because the Mössbauer spectra do not exhibit the effects of the internal magnetic field which would result from a paramagnetic system. In addition, the EPR results and the magnetic susceptibility data (159) show that these proteins are diamagnetic in the oxidized state.

If the iron sites are high-spin ferric (S = 5/2), then an exchange coupling mechanism is necessary to account for diamagnetism of the proteins in this oxidation state. Evidence for this exchange coupling between the iron sites will be given during the discussion of the reduced protein spectra. Since high-spin ferric is an S-state ion, the EFG which gives rise to the quadrupole splitting in the oxidized spectra must result from anisotropies in the ligand field surrounding the iron sites. In this case, the value of η for these spectra indicate that both axial and rhombic distortions are present in the ligand field. It is important that this be true since the $g = 1.94$ EPR signal of the reduced state can only be explained if these distortions are present. Some of the verification that these iron sites are both spin-coupled, high-spin ferric irons rests with the interpretation of the reduced protein spectra. Accordingly, we shall return to the discussion of the oxidized proteins after the presentation of the reduced protein data.

The Mössbauer spectra of spinach ferredoxin at 256 °K is shown in Fig. 9a; the solid line on these spectra is the result of computer-simulated Mössbauer spectra. A magnetic field of 46 kilogauss parallel to the gamma-ray direction was applied to this sample (Fig. 9b) in order to establish the sign of QS and the value of η. Inspection of the fourline, zero-field spectrum (Fig. 9a) reveals that this spectrum can be fit by two quadrupole pairs. The parameter for the computer simulated spectra shown in Fig. 9 are given below in Table 5.

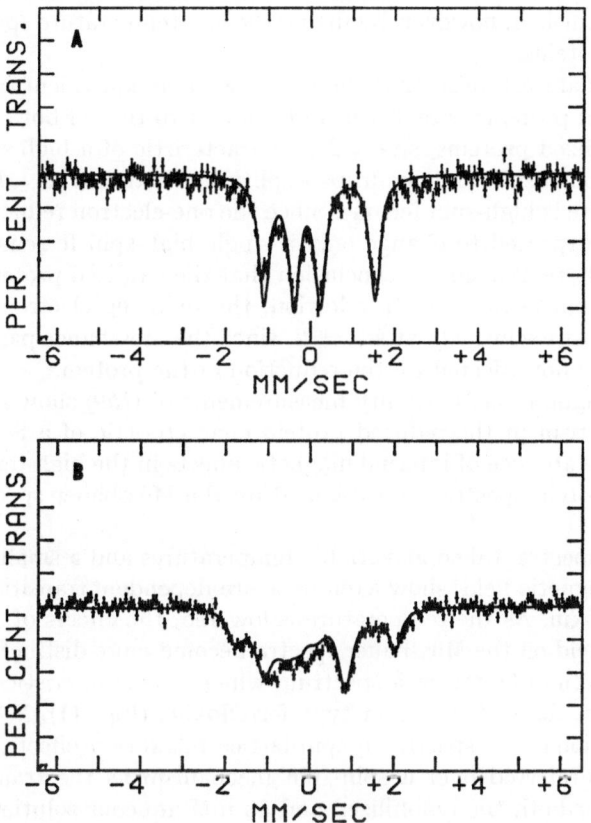

Fig. 9. Mössbauer spectra and computed Mössbauer spectra for reduced spinach ferredoxin at 256 °K. (a) Lyophilized spinach ferredoxin in zero magnetic field; (b) Lyophilized spinach ferredoxin with 46 kG magnetic field parallel to gamma-ray direction. Velocities relative to platinum source

The assignment of quadrupole pairs shown in Table 5 is the result of a trial-and-error approach to fit the high-field data with computer-simulated spectra. This approach establishes, unambiguously, the values for the isomer shift and magnitude of the quadrupole splitting shown for iron sites in Table 5. In addition, the sign of QS for iron site # 2 is determined with no assumptions in interpretation during curve fitting procedures. Noticing that the values of IS and QS for iron site # 1 are the same as for the sites in the oxidized proteins, we then assume that the value of η for iron site # 1 is the same in oxidized and reduced proteins. With this assumption, the value of η for iron site # 2 can be specified by the goodness of computer fits to the range, 0 to 0.5. The uncertainty in the value of η is diminished, however, by fitting the low-temperature spectra of the reduced proteins.

These data establish that there are two non-equivalent iron sites in the reduced proteins: site # 1 is quite similar to that of both iron atoms in the oxidized proteins, site # 2 is characteristic of a high-spin ferrous ion. The isomer shift and quadrupole splitting of site # 2 leave little doubt that this site is high-spin ferrous. Since the one-electron reduction of this protein is expected to change only a single high-spin ferrous ion, these data greatly reinforced the conclusion that the oxidized proteins contain two high-spin ferric ions. In addition, the reducing electron is seen to reside almost exclusively at site # 2, since the Mössbauer parameters of site # 1 are not affected by the reduction of the protein.

The magnetic susceptibility measurements of (159) show a molecular paramagnetism in the reduced protein characteristic of a $S = 1/2$ compound. The absence of internal magnetic effects in the high-temperature, reduced-protein spectra are explained by the Mössbauer spectra shown in Fig. 10.

These spectra, taken at variable temperatures and a small polarizing applied magnetic field, show a temperature-dependent transition for spinach ferredoxin. As the temperature is lowered, the effects of an internal magnetic field on the Mössbauer spectra become more distinct until they result at around 30 °K, in a spectrum which is characteristic of the low temperature data of the plant-type ferredoxins (Fig. 11). We attribute this transition in the spectra to spin-lattice relaxation effects. This conclusion is preferred over a spin-spin mechanism as the transition was identical for both the lyophilized and 10 mM aqueous solution samples. Thus, the variable temperature data for reduced spinach ferredoxin indicate that the electron-spin relaxation time is around 10^{-7} seconds at 50 °K. The temperature at which this transition in the Mössbauer spectra is half-complete is estimated to be the following: spinach ferredoxin, 50 K; parsley ferredoxin, 60 °K; adrenodoxin, putidaredoxin, Clostridium and Axotobacter iron-sulfur proteins, 100 °K.

The Mössbauer spectra of the reduced proteins at 4.2 °K are shown in Fig. 11 for 3.4 kilogauss applied field and in Fig. 12 for 46 kilogauss applied field. Since the spectra are so similar, we shall speak exclusively in terms of the spinach ferredoxin data. Fig. 13 is low-temperature spinach-ferredoxin spectra with computed fits superimposed. By assuming that the isomer shift and quadrupole parameters for the low temperature spectra are the same as for the high temperature spectra and then adjusting magnetic parameters by trial and error, we were able to obtain a set of "best fit" magnetic parameters for the low temperature spectra. The hyperfine constants for site # 1 which resulted from this approach were very close to those measured independently by *R. H. Sands, J. Fritz,* and *J. Fee* by ENDOR experiments *(160)*. Since hyperfine constants measured by ENDOR are more precise than those measured by Mössbauer spectroscopy, the ENDOR results were adopted for site # 1.

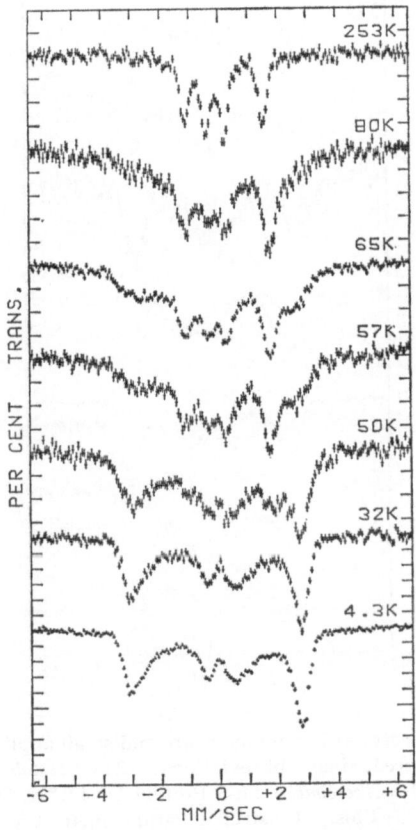

Fig. 10. Mössbauer spectra taken at various temperatures between 4.3 °K and 253 °K for lyophilized spinach ferredoxin with 580 G magnetic field applied parallel to the gamma-ray direction. Velocity scale relative to platinum-source

Using these "improved parameters" for site # 1, the trial and error approach was then resumed in order to find a best fit for the site # 2 parameters. Subsequently, the ENDOR values for the hyperfine interaction at site # 2 were also obtained by *Sands* and his co-workers (*160*). Since these values were also in agreement with our own, the final parameters for spinach ferredoxin shown in Table 6 incorporate the combined effort of ENDOR and Mössbauer results, although the ENDOR results give no

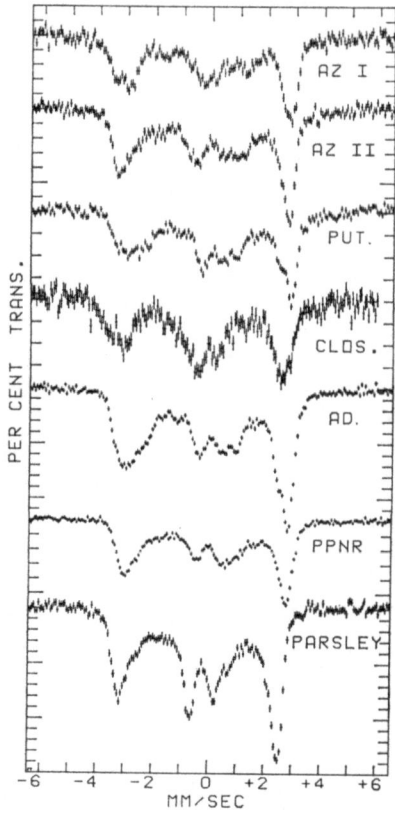

Fig. 11. Mössbauer spectra at low temperature and small applied magnetic field for reduced plant-type ferredoxins. Abbreviations: AZI = *Azotobacter* Fe-S Protein I, 4.2 °K, 1.15 kG; AZII = *Azotobacter* Fe-S Protein II, 4.2 °K, 300 G; Put. = Putidaredoxin, 4.6 °K, 580 G; Clos. = Clostridial Paramagnetic Protein, 4.7 °K, 3.4 kG; Ad. = Pig Adrenodoxin, lyophilized, 5.3 °K, 580 G; PPNR = Spinach Ferredoxin, lyophilized, 4.3 °K, 580 G; Parsley = Parsley Ferredoxin, 5.1 °K, 580 G. Applied magnetic field is parallel to gamma-ray direction. Velocities are relative to platinum source matrix

Table 6. *Mössbauer parameters for the low temperature reduced PPNR spectra*

	IS/Pt	QS (mm/S)	A_x	A_v	A_z	G_x	G_y	G_z
			(In electron gauss)					
Iron # 1	−0.10	+0.64 .5	− 17.8	− 18.6	− 15.1	1.89	1.96	2.05
Iron # 2	+0.19	− 2.68 .15	+5.0	+7.1	+12.5	1.89	1.96	2.05

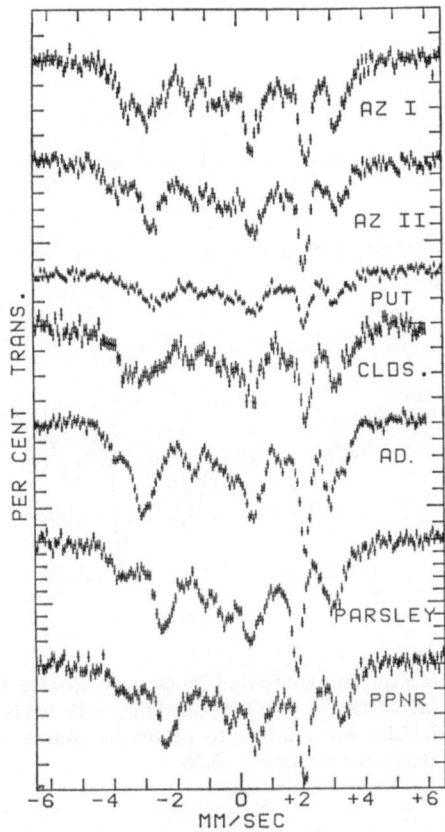

Fig. 12. Mössbauer spectra at low temperature and high applied magnetic field for reduced plant-type ferredoxins. Abbreviations: AZI = *Azotobacter* Fe-S Protein I, 4.2 °K, 46 kG; AZII = *Azotobacter* Fe-S Protein II, 4.2 °K, 46 kG; Put. = Putidaredoxin, 4.6 °K, 46 kG; Clos. = Clostridial Paramagnetic Protein, 4.2 °K, 46 kG; PPNR = Spinach Ferredoxin, lyophilized, 4.3 °K, 46 kG. Applied magnetic field is parallel to gamma-ray direction. Velocities are relative to platinum source matrix

information on the sign of the principle Ã tensor components. The spectra in Fig. 13 show the computed Mössbauer spectra which result from the parameters in Table 6.

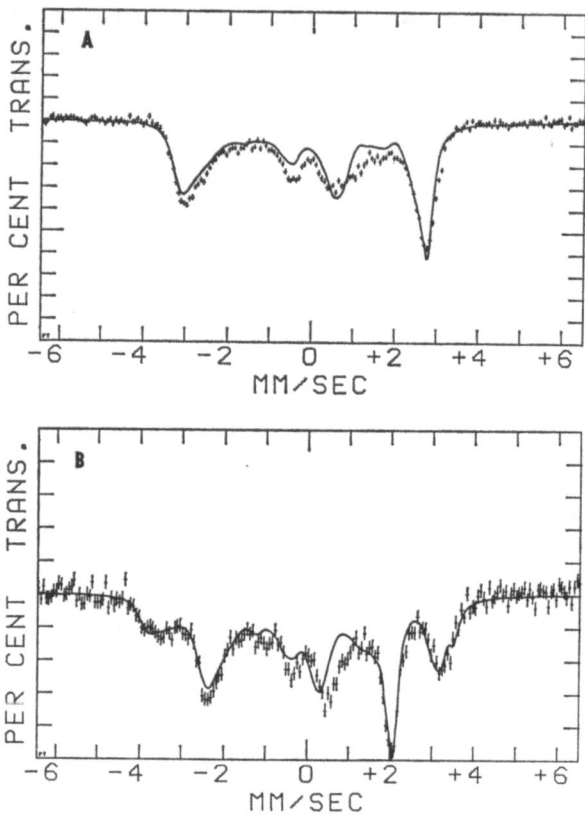

Fig. 13. Mössbauer spectra and computed Mössbauer spectra for reduced spinach ferredoxin at 4.3 °K. (a) 580 G; (b) 46 kG. Applied magnetic fields are parallel to gammaray direction. Velocities are relative to platinum source matrix. Boltzmann weighting factor for electronic states = − 0.26

In order to explain the spectra in Fig. 13 it is necessary to introduce another parameter into the discussion. Consider the effect of applying an external magnetic field to an S = 1/2 system. The effect of the field is to create two electron spin populations: one with spin parallel to the applied field and one with spin anti-parallel to the applied field. Further, these spin states will have different populations given by a Boltzmann factor. Note also that because the magnetic moment of the spin, with respect to the applied field, is reversed for the two spin states, the magnitude of the effective magnetic field at the nucleus differs for the two spin states by twice the amount of applied magnetic field. An applied magnetic field of around 30 kilogauss is necessary in order that the Mössbauer spectra of the two spin states become distinct. When the applied field is around 30 kilogauss, low temperatures of approximately 5 °K are needed to cause the differences in population of the two states to become measurable by Mössbauer spectroscopy. When the applied field is 46 kilogauss and the temperature is 4.2 °K, as is the case in Fig. 13, both of these criteria are met. Therefore, Fig. 13 contains Boltzmann parameters, 0.26 and 1.0 for the populations of the two spin states for the resultant spin one-half system of the reduced protein complex.

As added evidence for our confidence in the parameters shown in Table 6, the zero applied field spectra taken at low temperatures are shown in Fig. 13. Since the A-values for site # 1 are almost isotropic, it is expected that the absorption peaks from this site would dominate the Mössbauer spectra in both zero and applied magnetic field. Comparison of Fig. 14 and Fig. 3 reveals that the absorption in these spectra at − 6 mm/S results from an isotropic hyperfine interaction of about − 17 gauss at one of the iron sites in the reduced proteins. The anisotropic hyperfine interaction at site # 2 results in a broad, unresolved absorption which accounts for the difference in shape between the spectra.

The Mössbauer spectra for these proteins are consistent with the "spin-coupled" model proposed by *Gibson et al.* (*148*) for the active site of these proteins. In the next section we shall discuss this model in detail.

The iron atoms in the oxidized protein are high-spin ferric ($^6S_{5/2}$) ions, exchange coupled to give a resultant spin-zero complex. Upon reduction, one of the iron atoms changes to the high-spin ferrous state (S = 2). The exchange coupling for this protein oxidation state gives a resultant spin of one-half. *Lewis et al.* (*160*), *Khedekar et al.* (*161*) and *Gerloch et al.* (*162*) have observed a similar exchange-coupling mechanism in a number of Schiff's base iron salts. In every case in which the exchange coupling constant was negative (antiferromagnetic), the structure of the salt is as shown below:

$$R_1 \diagdown O \diagup R_2$$
$$Fe Fe$$
$$R_3 \diagup O \diagdown R_4$$

where the R's refer to the Schiff-base ligands. If this situation is analogous to that in the plant-type ferredoxins, then we may assume that the role of the labile sulfur in these proteins is to bridge the iron atoms in an analogous fashion and thus promote the exchange coupling interaction.

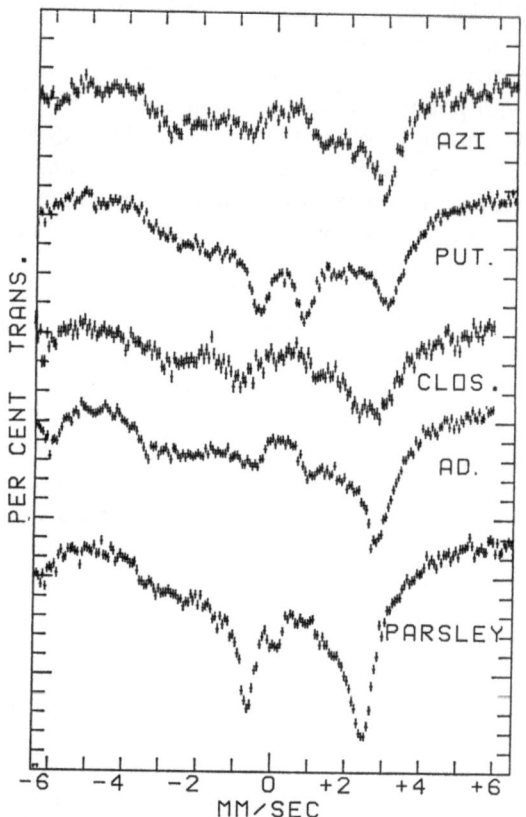

Fig. 14. Mössbauer spectra of reduced plant-type ferredoxins in zero magnetic field at low temperature. Abbreviations: AZI = *Azotobacter* Fe-S Protein I, 4.2 °K; Put. = Putidaredoxin, 4.2 °K; Clos. = Clostridial Paramagnetic Protein, 4.2 °K; Ad. = Adrenodoxin, 4.2 °K; Parsley = Parsley Ferredoxin, 4.6 °K. Velocity scale is relative to source in platinum matrix

The $g = 1.94$ EPR signal of the reduced proteins must be explained by any model for their active site. Using subscript 1 to specify the ferric-iron site and subscript 2 the ferrous-iron site, the "spin-coupled" model explains this EPR signal in the following way. The electron magnetic moments ($S_1 = 5/2$ and $S_2 = 2$) are coupled to form a resultant spin S as shown below.

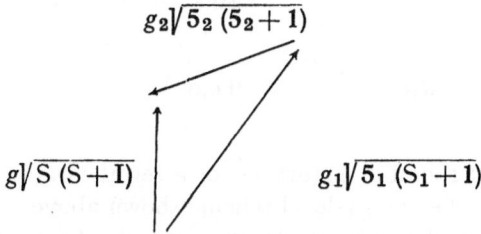

From the law of the cosines, the g-value for an $S = 1/2$ system is given by the following expression:

$$g = (7g_1 - 4g_2) \, / \, 3$$

Since g_1 arises from an S-state ion, spin-orbit interactions are not allowed to first order (163) and g_1 can therefore be assumed to be isotropic. It is assumed to be 2.019 in accord with the measurements of *Title* (166). With this assumption, the g-values for the ferrous iron can be derived using the above equation and the measured g-values for the proteins (Table 6). For spinach ferredoxin, these calculated values are $g_{2x} = 2.12$, $g_{2y} = 2.07$, and $g_{2z} = 2.00$.

In the high-spin ferrous ion, spin-orbit interactions mix the ground state wave functions with the excited states. If the ground state is assumed to have d_{z^2} symmetry, then the following expressions apply for an ion in a crystal field with both rhombic and axial distortions (*Edwards et al.* (70).

$$g_{2x} = g_e + 6\lambda/\Delta_{yz}$$
$$g_{2y} = g_e + 6\lambda/\Delta_{xz}$$
$$g_{2z} = g_e$$

where λ is the spin-orbit coupling constant in the interaction $\lambda \overline{L} \cdot \overline{S}$. Δ_{xz} and Δ_{yz} are the energy gaps to the excited states having d_{xz} and d_{yz} symmetries, respectively. These expressions are derived by assuming that the electronic ground state is equivalent to a hole with spin $= 2$ in a d_{z^2} orbital. λ can be estimated to be 80 cm^{-1} by taking into account the effects of covalency on other high-spin ferrous ions (70). With the above assumptions, one can derive the following energy level scheme for the high-spin ferrous ion.

$$d_{yz} \quad \underline{\qquad}$$

$$d_{xz} \quad \underline{\qquad} \quad 6900 \text{ cm}^{-1}$$

$$d_{yz} \quad \underline{\qquad} \quad 4000 \text{ cm}^{-1}$$

$$d_{x^2-y^2} \quad \underline{\qquad}$$

$$d_{z^2} \quad \underline{\qquad} \quad 0 \text{ cm}^{-1}$$

Both axial and rhombic distortions of a tetrahedral ligand field are necessary to cause the energy-level scheme shown above.

The energy levels shown for the ferrous-iron site of the reduced protein is based on the assumption that the electron pair in the d-orbital system of this ion occupies a d_{z^2} orbital. The proof of this assumption lies in the values of the derived parameters for the low temperature spectra of the reduced proteins. Consider first the parameter, QS. The only d-orbitals which give negative values for QS are d_{z^2}, d_{xz} and d_{yz}. A large negative, value -2.63 mm/S, for site # 2 (Table 6) agrees well with that calculated for a single electron in a d_{z^2} orbital. The experimental value of η is close to zero for the ferrous iron. This value is inconsistent with the theoretical values of η for d_{xz} and d_{yz} orbital density. In addition, the magnitude of the measured value of QS (-2.63 mm/S) is very close to that predicted for a d_{z^2} electron: -3 mm/S (167).

Other Mössbauer data which indicate that the model is correct are the measured a-values for the low-temperature, reduced-protein spectra. The measured a-values for the ferric iron (Table 6) are close to isotropic with an average value of -17 gauss. Remembering that this a-value is calculated for an electron spin $= 1/2$ situation, we now recalculate the a-value for the ferric site in terms of the 5/2 spin present at this site. For the ferric site in the spin-coupled model,

$$a_1 \, (S_1 = 5/2) - 7/15 < a_1 >_{\text{measured}} = -8 \text{ gauss}$$

In high-spin ferric iron this a-value is the result of the Fermi contact interaction alone. Hence, this a-value comprises an experimental determination of the Fermi contact interaction of the ferric iron in this protein.

The value of $-2g_N\beta\beta_N <r^{-3}>$ is, by the above procedure, equal to -1.6 gauss. We now apply this constant to the calculation of the a-values for the ferrous iron. The Fermi contact interaction at the ferrous site is approximated by assuming that $-2g_N\beta\beta_N <r^{-3}>$ for this site equals -1.6 gauss times 0.87. A value of 0.35 is assumed for K, thus scaling the Fermi contact interaction to the dipolar interaction. These values are then entered using the data for the dipolar part of the hyperfine interaction. Using the orbital scheme the a-values for the ferrous iron are then

$$a_x = -4.5 \text{ gauss}$$
$$a_y = -4.5 \text{ gauss}$$
$$a_z = -8 \text{ gauss}$$

Following a procedure analogous to that used before, a set of a-values are computed which correspond to those measured by Mössbauer spectroscopy for the ferrous iron. Table 7 shows these completed and measured a-values for the ferrous site.

Table 7. *A-Values for ferrous iron*

Computed	Measured
$a_x = +\ 6.6$ gauss	$a_x = +\ 5.0$ gauss
$a_y = +\ 6.6$ gauss	$a_y = +\ 7.1$ gauss
$a_z = +11.8$ gauss	$a_z = +12.5$ gauss

The agreement in the values of Table 7 not only indicate the validity of our assumptions regarding the hyperfine interaction of the ferrous iron, but also comprises a rigorous test for the model as a whole, since the presence of positive a-values for iron with magnitudes shown in Table 6 necessitates an exchange coupling mechanism.

In the preceding section we presented the experimental evidence in support of the "spin-coupled" model proposed by *Gibson et al.* (*148*) and *Thornley et al.* (*150*) for the plant-type ferredoxins. However, the "spin-coupled" model does not provide a spatial or configurational model for the active center. Therefore we proceed to a more detailed analysis with the goal of asserting a proper chemical and structural model of the active center. The following properties of the active site of these proteins are well-substantiated experimentally.

1. The active center of the oxidized plant-type ferredoxins contains two iron atoms with almost identical electronic environments at the nuclei. These irons are high-spin ferric ($S = 5/2$), spin-coupled to give a resultant diamagnetism for the complex.

2. In the reduced state, the active center contains one high-spin ferric state spin-coupled to one high ferrous state ($S = 2$) to give a resultant $S = 1/2$ complex.

3. The ligand symmetry around both iron atoms is tetrahedral, but with both axial and rhombic distortions. This basic symmetry is not affected by reduction of the proteins.

4. The active center of the plant-type ferredoxins is nearly identical in every protein studied. The only differences in this active center are the presence and magnitude of the rhombic distortion of the symmetry for the ferrous iron in the reduced proteins.

In addition, the two-iron Schiff's base compounds studied by *Lewis et al.* (*160—162*) have magnetic properties which indicate a structure which may be similar to that in the active centers of the plant-type ferredoxins. The following arguments set forth criteria on which to base any model for the active site:

1. The iron atoms have been shown to be exchange coupled through a superexchange mechanism. Thus, they are connected by a bridging ligand which, in view of the arguments in the previous section and the elemental composition of the holoprotein, is almost likely a sulfur atom. This bridging ligand (sulfur) can, however, be either cysteine sulfur or "labile sulfur".

2. The ^{33}S EPR experiments show that the "labile sulfur" atoms are in the active site. The magnitude of the sulfur hyperfine constants indicate that the "labile sulfur" is bonded to the iron. In view of the amino acid compositions of these proteins, the "labile sulfur" is either the bridging ligands for the iron atoms or part of a persulfide ligand to the iron atoms.

Thus, the following persulfide structures are consistent with the above criteria:

L = Ligand from amino acid side chains.

Although no direct evidence has allowed a decision between these structures, we feel that both the structures shown above are doubtful because: a) the persulfide bonds are higher energy than sulfur-iron bonds, and b) these structures do not promote the similarities which are observed in the Mössbauer spectra of the seven proteins in this study.

Note that in the structure in Fig. 15, there are six ligands: two "labile sulfur", iron-bridging ligands and four cysteinyl-sulfur ligands.

The tetrahedral ligand symmetry of this model is distorted by the difference in character between the "labile" and cysteinyl sulfur atoms and by the position of the iron atoms themselves.

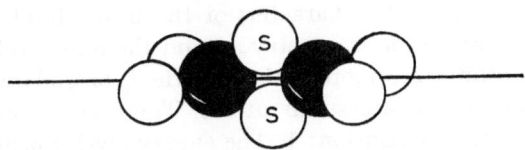

Fig. 15. A Model of the Active Center of the Fe-S Plant-Type Ferredoxins

In crystal field theory calculations the direction of the axial distortion is along the z-axis. Therefore, the d_z2 orbitals in iron atoms in Fig. 15 are along the line adjoining the two iron atoms. Remembering that the d_z2 orbital lies lowest in this symmetry, the effect of reducing the complex is to add electron density to the d_{z^2} orbitals of the iron atoms. Since the d_{z^2} iron-orbitals in Fig. 15 overlap, this structure results in an electron repulsion term between the iron atoms which increases as the iron atoms in these proteins are reduced. Thus, the negative reduction potentials (Table 1) of the plant-type ferredoxins can be accounted for by this model.

The protein sequence data in Table 2 show that the cysteine residues in all the proteins occur in identical positions (*18, 39, 44, 47, 77*) in the sequence. Thus, the ligand field produced by the cysteinyl-sulfur atoms is not likely to be different among these proteins unless there is a difference in protein conformation which causes a displacement in one or more of the cysteinyl sulfur atoms. Note that a displacement of *any* cysteinyl sulfur atom in the model in Fig. 15 results in rhombic distortion at the iron to which it is ligated. Since, according to the "spin-coupled" model, this rhombic distortion will manifest itself in the difference between g_x and g_x for a particular protein, the EPR data in Table 1 provide a measure of the rhombic distortion around the ferrous iron in the reduced proteins. In particular, the g-values of adrenodoxin are axially symmetric while the g-values of spinach ferredoxin show a rhombic distortion. Thus, the observation of *Kimura et al.* (*168*) that adrenodoxin and spinach ferredoxin have different protein conformations is consistent with the prediction of the above model.

The "spin-coupled" model also predicts a constant value of g_z, numerically 2.04. Inspection of Table 1 shows some deviation from $g_z = 2.04$ for the plant-type ferredoxins. With respect to the model shown

41

in Fig. 15, we must invoke large strains on the cysteine sulfurs around the reduced protein ferric site in order to account for the deviations as the "spin-coupled" model attributes the value of g_z mainly to the g-value of the ferric site.

Assuming that the structure shown in Fig. 15 is valid, one can draw some conclusions as to the characters of the iron-orbitals in the "spin-coupled" model. Since the symmetry around the iron is tetrahedral, the d_{z^2} and $d_{x^2-y^2}$ orbital are more ionic than the t_g orbitals which must be covalent as the ligands are sulfur atoms. There are several important consequences of this conclusion: 1. The energy level scheme is based on crystal-field approximations and therefore can be considerably in error. 2. The reducing electrons will occupy an ionic orbitals (d_{z^2}); thus, the reduction potential of the plant-type ferredoxins is justifiably attributed to electron repulsion between the d_{z^2} orbitals of the two iron atoms. 3. Covalency of the t_{2g} iron orbitals with sulfur implies that the Mössbauer spectra of these proteins will be sensitive to ligand changes between the members of the plant-type ferredoxins. That is, the substitution of tyrosine or histidine for cysteine as a ligand is certain to cause a change in isomer shift which is not observed for these proteins.

In fact, the similarities in the Mössbauer effect parameters impose tight constraints on the freedom to choose ligands for this complex. Thus, the suggestion by *Yang* and *Huennekens* (*169*) that iron-sulfur complex involves octahedral hydroxy ligands including tyrosine is not applicable on two counts: 1. the ligand symmetry is tetrahedral and, 2. the positions of the tyrosine residues are not constant throughout the sequence of the plant-type ferredoxins.

The acidity of these proteins implies that the amino acids which occur in areas of the sequence with a preponderance of glutamic acid, aspartic acid and glutamine will not be free to ligate to the iron-sulfur complex, as they will be drawn out to the periphery of the protein conformation. Thus, consideration of the similarities in the Mössbauer spectra and inspection of the amino acid sequences and composition (170, *171*) imply that cysteine is the most probable ligand to the iron-sulfur complex, and that the structure shown in Fig. 15 is valid.

B. Bacterial-Type Iron-Sulfur Proteins

The bacterial-type iron-sulfur proteins all contain larger amounts of iron and labile sulfide than the plant-type iron-sulfur proteins; best estimates for the iron and labile sulfide content being 8 Fe and 8 S per protein molecule (*172, 173*) for these ferredoxins from *Clostridium* and from *Chromatium*. Although these proteins have large amounts of Fe and S, the molecular weights are less than the molecular weights of the plant-type iron-sulfur proteins, typically about 6000 for the bacterial-type ferredoxin from *Clostridium pasteuranium* and *Clostridium acidi-urici.* The biochemistry of these materials and the physiological function, where known, are discussed in several review articles previously mentioned (*129, 133,.135*).

The first Mössbauer spectroscopic studies on this class of iron-sulfur proteins were carried out by *Blomstrom et al.* (*174*) who reported only spectra taken on the oxidized form of the protein from *Clostridium pasteuranium.* Evidence in the form of two partially resolved quadrupole pairs was presented to support two distinct environments for the iron in the oxidized form of the protein, the ratio of 5:2 was suggested in keeping with the then thought number of Fe atoms per protein molecule. This assignment of the number of Fe atom per site, of course, rests on the assumption of equal Lamb-Mössbauer recoil-free fractions for the two sites.

The ferredoxin from *Chromatium* has also been examined by Möss-bauer techniques (*152*), and in this case both the oxidized and the dithio-nite-reduced forms of the protein were examined, although there is now some doubt as to the validity of the data for the reduced form. In the oxidized form, there are again two Fe sites in about the ratio 2:1. The quadrupole splittings observed (1.32 and 0.66 mm/S at 4 °K) are within the range for high-spin Fe(III) biomolecules; in addition, there is no evidence for a magnetic hyperfine interaction even at 4 °K. No studies of this material have ever been made under conditions of an external magnetic field. When the *Chromatium* bacterial-type iron-sulfur protein was reduced with sodium dithionite, a pair of quadrupole lines with a splitting of 2.9 mm/S appeared, and accounted for about 1/6th of the total Mössbauer absorption; later studies (*175*) with *Clostridial acidiurici* point out that the reducing conditions and buffer concentration used in the earlier work may have not preserved the reduced form of the protein, but may be indicative of denatured protein; these studies with the reduced form of the protein certainly bear repeating.

The absence of magnetic hyperfine interactions in the oxidized forms of the bacterial-type iron-sulfur proteins does point to the possibility of spin-pairing within the molecule; this feature is also shown more clearly by the reported diamagnetism and by the effects of the spin-coupling on line positions in high resolution proton magnetic resonance in the studies by *Poe et al.* (*176*). The fact that the bacterical-type iron-sulfur proteins act as *two* electron-transfer agents in contrast to the single electron tranferred in the plant-type iron-sulfur proteins and the discovery that there are *two* paramagnetic centers, each of the $g = 1.94$ variety of *Orme-Johnson* and *Beinert* (*177*) for *Clostridium pasteuranium* (and for the conjugated iron-sulfur protein, xanthine oxidase) indicates the complexity of these materials. Further study of Mössbauer spectroscopy in conjunction with other resonance techniques is needed to clarify this situation particularly keeping in mind the fact that the Fe-S complex can only be removed (and reconstituted) as a complete unit (178).

VI. Hemerythrin

Hemerythrin is a non-heme iron-containing protein which is distinct from the iron-sulfur proteins previously discussed. This protein serves as a reversible oxygen carrier in red cells of some brachipods and sipunculids. (179—181) Hemerythrin can be dissociated into eight subunits each of molecular weight 13,500 which contain two Fe atoms and can bind reversibly one O_2 molecule (182). There is no "acid-labile" sulfide present in the molecule. The determination of the oxidation state of the iron and the mechanism of reversible oxygenation has been a goal for some time; chemical studies by *Klotz* and *Klotz* (183) which suggest a ferric peroxy form of binding have been questioned by other workers (184, 185). Thus it is important to apply physical techniques, particularly magnetic susceptibility measurements and Mössbauer spectroscopy in order to resolve this question.

Magnetic susceptibility measurements by *York* and *Moss* (186) over the temperature range from 14 °K to 77 °K show the absence of any paramagnetism for the F⁻, N_3, OCN⁻, SCN⁻, CN⁻ derivatives of the oxidized protein, and for the deoxygenated and oxygenated forms as well. Room temperature magnetic susceptibility measurements (187) show a paramagnetism close to the value expected for four unpaired electrons (S = 2) for deoxygenated hemerythrin and an absence of paramagnetism for the oxidized derivatives. A paramagnetism corresponding to a single electron per Fe atom is found for the oxygenated form. These quantitative magnetic susceptibility measurements are in agreement with relative measurements made by *Okamura* and co-workers (181).

Mössbauer spectroscopic measurements on hemerythrin were first performed by *Gilchrist* (188) and later by *York* and *Bearden* (186), *Okamura et al.* (188), and by *Garbett et al.* (189). The spectral results as obtained initially are now agreed to by all researchers and can be summarized as follows. The oxidized derivatives of hemerythrin display a single quadrupole pair of lines with a quadrupole splitting of about 1.8 mm/S with a very slight temperature dependence. This wide splitting is similar to the splitting observed in methemoglobin and metmyoglobin and is considerably wider than observed for ionic Fe(III) model compounds. This point has been well discussed under the section on the Mössbauer spectroscopy of hemoproteins; it is only necessary to recall that this wide pair is attributable to the large distortions of the Fe(III) electronic configuration by the porphyrin ring and protein-binding ligand. Such a distorted configuration must be present in this non-porphyrin moiety in hemerythrin. There is no evidence for a magnetic hyperfine interaction down to 4 °K and in applied magnetic fields of several kilogauss.

For the deoxygenated form of hemerythrin, a single quadrupole pair is again present, but with a much wider quadrupole splitting (2.8 mm/S) indicative of a high-spin Fe(II) electronic configuration. No hyperfine interaction is present down to 4 °K. In each form of hemerythrin discussed so far, all of the Fe is present in a single site in each form.

The oxygenated form of the protein has a Mössbauer spectra consisting of two pairs of quadrupole lines indicating *two* distinct Fe sites. Again there is no evidence of hyperfine interaction down to 4 °K. The attribution of one pair of lines to impurities by *Okamura et al. (181)* has now been retracted by this group of researchers *(189)*.

Garbett et al. (189) and *Rill* and *Klotz (190)* present a mechanism supporting the peroxo-bridge arrangement for the oxygen-bound form of hemerythrin; the difficulty of explaining the *two* distinct sites of Fe rests on subsequent reaction mechanisms which are not easily demonstrable. The most important conclusion based on Mössbauer evidence is the absence of hyperfine interactions for any of the forms of the protein. The absence of paramagnetism for the oxidized and oxygenated derivatives indicate that there is a "spin-paired" diamagnetic molecule.

The two environments indicated by the Mössbauer data and the fact that during the reaction of tetranitromethane with tyrosine residues only one of the two atoms per protein in subunit was released *(190)* points out the possibility that O_2 binding to a single iron site must still be considered a possibility. The recent work on O_2 binding to Bis [bis(diphenylphosphino) ethane] iridium(I) Hexafluorophosphate *(191)* and a series of other papers on O_2-binding to binuclear complexes *(192, 193)* may be relevant to the O_2 binding of hemerythrin. Further studies by magnetic susceptibility with the full temperature range available in a single instrument and by Mössbauer spectroscopy with the availability of a large external magnetic field, and the possibility of ^{57}Fe enrichment would be most helpful in this problem.

Summary

Resonance spectroscopies have an important role in determining electronic configurations and protein conformations. High resolution proton magnetic resonance may provide information on conformations in solution, an essential adjunct to studies by x-ray crystallography. Mössbauer spectroscopy is most valuable for materials which are not so far amenable to x-ray study; for example, studies of iron-sulfur proteins and other nonheme proteins.

For proteins whose structure is known by means of crystallographic studies, Mössbauer spectroscopy and studies by electron paramagnetic resonance afford an opportunity to determine the detailed electronic configurations, a necessary step towards the chemical basis of protein function.

Acknowledgements

The authors are in debt to their many colleagues and co-workers who have labored in the applications of resonance spectroscopy to biological problems. We particularly acknowledge the discussion of these matters with Professors *D. I. Arnon, Helmut Beinert, A. Ehrenberg, I. C. Gunsalus, M. D. Kamen, J. B. Neilands, L. E. Orgel,* and *J. C. Rabinowitz* and with Drs. *R. G. Bartsch, B. B. Buchanan, W. D. Phillips,* and *I. Salmeen.*

The report-in-depth on Mössbauer spectroscopy of the iron-sulfur proteins includes collaborative research, also published elsewhere, with *W. H. Orme-Johnson, Graham Palmer, R. H. Sands,* and *I. Salmeen.*

This work has been supported through grants-in-aid from the National Science Foundation (GB-13585) and from the U.S. Atomic Energy Commission through Donner Laboratory. The first author (AJB) has been partially supported by a Public Health Service Research Career Development Award (1-K4-GM-24,494-01) from the Institute of General Medical Studies. The second author (WRD) has been supported in part by a Biochemistry Training Grant from the National Institutes of Health through the Department of Chemistry, University of California, San Diego.

References

1. *McDonald, C. C., Phillips, W. D.:* In: Magnetic Resonance in Biological Systems, p. 3; edited by *Ehrenberg, A., Malmström, B. G., Vänngård, T.* Oxford: Pergamon Press 1967.
2. — — J. Am. Chem. Soc. *97*, 1513 (1969).
3. *Shulman, R. G., Wuthrich, K., Yamane, T., Antonini, E., Brunoni, M.:* Proc. Natl. Acad. Sci. U.S. *63*, 623 (1969).
4. *Beinert, H., Orme-Johnson, W. H.:* Annals of the New York Academy of Sciences; Electronic Aspects of Biochemistry, *158*, Art. 1, p. 336 (1969).
5. — — In: Magnetic Resonance in Biological Systems, p. 221; edited by *Ehrenberg, A., Malmström, B. G., Vänngård, T.* Oxford: Pergamon Press 1967.
6. *Palmer, G., Brintzinger, H., Estabrook, R. W., Sands, R. H.:* In: Magnetic Resonance in Biological Systems, p. 173; edited by *Ehrenberg, A., Malmström, B. G., Vänngård, T.* Oxford: Pergamon Press 1967.

7. *Hamilton, C. L., McConnell, H. M.:* In: Structural Chemistry and Molecular Biology, p. 115; edited by *Rich, A., Davidson, N.:* San Francisco: W. H. Freeman 1968.
8. *McConnell, H. M., Deal, W., Ogata, R. T.:* Biochemistry *8*, 2580 (1969).
9. *Hubbell, W. L., McConnell, H. M.:* Proc. Natl. Acad. Sci. U. S. *63*, 16 (1969).
10. *Phillips, W. D., Knight, E., Jr., Blomstrom, D. C.:* In: Non-Heme Iron Proteins: Role in Energy Conversion, p. 69; edited by *San Pietro, A.* Yellow Springs, Ohio: Antioch Press 1965.
11. *Bearden, A. J., Moss, T. H., Bartsch, R. G., Cusanovich, M. A.:* In: Non-Heme Iron Proteins: Role in Energy Conversion, p. 87; edited by *San Pietro, A.* Yellow Springs, Ohio: Antioch Press 1965.
12. *Bearden, A. J., Moss, T. H.:* In: Magnetic Resonance in Biological Systems, p. 391; edited by *Ehrenberg, A., Malmström, B. G., Vänngård, T.* Oxford: Pergamon Press 1967. Also Ref. 196.
13. *Debrunner, P., Tsibris, J. C. M., Münck, E.* (editors): Mössbauer Spectroscopy in Biological System. University of Illinois Bulletin Urbana, 1969.
14. *Wertheim, G. K.:* Mössbauer Effect: Principles and Applications. New York: Academic Press 1964.
15. *Goldanskii, V. I., Herber, R. H.:* Chemical Applications of Mössbauer Spectroscopy. New York: Academic Press 1968.
16. *Danon, J.:* In: Physical Methods in Advanced Inorganic Chemistry, p. 380; edited by *Hill, H. A. O., Day, P.* New York: John Wiley/Interscience 1968.
17. *Herber, R. H.:* Ann. Rev. Phys. Chem., *17*, 261 (1966).
18. *Shirley, D. A.:* Ann. Rev. Phys. Chem. *20*, 25 (1969).
19. *Moss, T. H., Bearden, A. J., Bartsch, R. G., Cusanovich, M. A., San Pietro, A.:* Biochemistry *7*, 1591 (1968).
20. *Bearden, A. J., Moss, T. H., Bartsch, R. G., Cusanovich, M. A.:* In: Non-Heme Iron Proteins: Role in Energy Conversion, edited by *San Pietro, A.,* Yellow Springs, Ohio: Antioch Press 1965.
21. *Lang, G., Marshall, W.:* Proc. Phys. Soc. (London) *87*, 3 (1966).
22. *Gonser, U., Grant, R. W., Kregzde, J.:* Science *143*, 680 (1964).
23. *Moss, T. H., Ehrenberg, A., Bearden, A. J.:* Biochemistry *8*, 4159 (1969).
24. *Caughey, W. S., Fujimoto, W. Y., Bearden, A. J., Moss, T. H.:* Biochemistry *5*, 1255 (1966).
25. *Dunham, W. R., Bearden, A. J.:* unpublished data.
26. *Ehrenberg, A.:* In: Hemes and Hemoproteins, p. 133; edited by *Chance, B., Estabrook, R. W., Yonetani, T.* New York: Academic Press 1966.
27. *George, P., Beetlestone, J., Griffith, J. S.:* In: Hematin Enzymes. Intern. Union Biochem. Symp. Ser. *19*, 105 (1961).
28. *Yonetani, T., Ehrenberg, A.:* In: Magnetic Resonance in Biological Systems, p. 151; edited by *Ehrenberg, A., Malmström, B., Vänngård, T.* Oxford: Pergamon Press 1967.
29. *Yonetani, T.:* In: Methods of Enzymology: Oxidations and Phosphorylations, p. 332; edited by *Estabrook, R. W., Pullman, M. E.* New York: Academic Press 1967.
30. *Spiro, T. G., Saltman, P.:* Struct. Bonding *6*, 116 (1969).
31. *Wickman, H. H., Klein, M. P., Shirley, D. A.:* Phys. Rev. *152*, 345 (1966).
32. *Wickman, H. H.:* Mössbauer Effect Method. *2*, 39 (1966).
33. *Klein, M. P.:* In: Magnetic Resonance in Biological Systems, p. 407; edited by *Ehrenberg, A., Malmström, B. G., Vänngård, T.* Oxford: Pergamon Press 1967.

A. J. Bearden and W. R. Dunham

34. *Nath, A., Harpold, M., Klein, M. P., Kundig, W.:* Chem. Phys. Letters 2, 471 (1968).
35. — — *Klein, M. P., Kundig, W., Lichtenstein, B.:* Radiation Effects 2, 211 (1970).
36. *Wertheim, G. K., Herber, R. H.:* J. Chem. Phys. 38, 2106 (1963).
37. *Danon, J.:* Lectures on the Mössbauer Effect. New York: Gordon and Breach 1968.
38. *Kittel, C.:* Quantum Theory of Solids. New York: John Wiley and Sons, third edition.
39. *Shulman, R. G., Sugano, S.:* J. Chem. Phys. 42, 39 (1965).
40. *Fluck, E., Kerler, W., Neuwirth, W.:* Angew. Chem. 2, 277 (1963).
41. *Muir, A. H., Jr., Ando, K. J., Coogan, H. M.:* Mössbauer Effect Data Index 1958—65. New York: John Wiley/Interscience 1966.
42. *Walker, L. R., Wertheim, G. K., Jaccarino, V.:* Phys. Rev. Letters 6, 98 (1961).
43. *Goldanskii, V. I., Makarov, E. F., Stukan, R. A.:* Teoriya i Eksperim. Khim. Akad. Nauk Ukr. SSR. 2, 504 (1966).
44. *Kistner, O. C., Sunyar, A. W.:* Phys. Rev. Letters 4, 412 (1960).
45. *Grant, R. W.:* In: Mössbauer Effect Methodology, Vol. 2, p. 23; edited by *Gruverman, I.* New York: Plenum Press 1966.
46. *Collins, R. L.:* J. Chem. Phys. 42, 1072 (1965).
47. *Johnson, C. E.:* In: Magnetic Resonance in Biological Systems, p. 405; edited by *Ehrenberg, A., Malmström, B. G., Vänngård, T.* Oxford: Pergamon Press 1967.
48. *Cohen, S. G., Gielen, P., Kaplow, R.:* Phys. Rev. 141, 423 (1966).
49. *Carrington, A., McLachlan, A. D.:* Introduction to Magnetic Resonance. New York: Harper and Row 1967.
50. *Abragam, A.:* The Principles of Nuclear Magnetism. Oxford: University Press 1963.
51. — — *Pryce, M. H. L.:* Proc. Roy. Soc. (London) A 205, 135 (1951).
52. *Frauenfelder, H.:* The Mössbauer Effect, p. 4; New York: Benjamin 1962.
53 *Goldanskii, V. I., Makarov, E. F.:* In: Chemical Applications of Mössbauer: Spectroscopy, p. 36; edited by *Goldanskii, V. I., Herber, R. H.* New York Academic Press 1968.
54. *Mullen, J. G.:* Phys. Letters 15, 15 (1965).
55. *Dash, J. G.:* In: Mössbauer Effect Methodology, Vol. 1, p. 107; edited by *Gruverman, I.* New York: Plenum Press 1965.
56. *O'Connor, D. A., Black, P. J.:* Proc. Phys. Soc. (London) 83, 941 (1964).
57. *Goldanskii, V. I., Karyagin, S. V., Makarov, E. F., Khrapov, V. V.:* Dubna Conference on Mössbauer Effect, July, 1962, p. 189 (These conference proceedings have been published by Consultants Bureau, New York, 1963).
58. *Orgel, L. E.:* An Introduction to Transition-Metal Chemistry Ligand-Field Theory. New York: John Wiley 1966, Second Edition.
59. *Ballhausen, K. J.:* Introduction to Ligand Field Theory. New York: McGraw-Hill 1962.
60. *Harris, G.:* J. Chem. Phys. 48, 2191 (1968).
61. *Loew, G. H., Ake, R. L.:* J. Chem. Phys. 51, 3143 (1969).
62. *Jesson, J. P., Trofimenko, S., Eaton, D. R.:* J. Am. Chem. Soc. 89, 3158 (1967).
63. — — *Weiher, J. F.:* J. Chem. Phys. 46, 1995 (1967).
64. *Robinson, A. B., Moss, T. H.:* Inorg. Chem. 7, 1692 (1968).
65. *Jørgensen, C. K.:* Struct. Bonding 1, 3 (1966).
66. *Golding, R. M., Jackson, F., Sinn, E.:* Theoret. Chim. Acta (Berlin) 15, 123 (1969).

48

67. *Burbridge, C. D., Goodgame, D. M. L., Goodgame, M.:* J. Chem. Soc. (A) 349 (1967).
68. *Phillips, C. S. G., Williams, R. J. P.:* Inorganic Chemistry. Oxford: University Press 1965.
69. *Ingalls, R.:* Phys. Rev. *128*, 1155 (1962).
70. *Edwards, P. R., Johnson, C. E., Williams, R. J. P.:* J. Chem. Phys. *47*, 2074 (1967).
71. — — J. Chem. Phys. *49*, 211 (1968).
72. *Maeda, Y.:* J. Phys. Soc. (Japan) *24*, 151 (1968).
73. *Shulman, R. G., Wertheim, G. K.:* Rev. Mod. Phys. *36*, 457 (1964).
74. *Gonser, U., Grant, R. W.:* Biophys. J. *5*, 823 (1965).
75. *Bearden, A. J., Moss, T. H., Caughey, W. S., Beaudreau, C. A.:* Proc. Natl. Acad. Sci. U.S. *53*, 1246 (1965).
76. *Epstein, L. M., Straub, D. K., Maricondi, C.:* Inorg. Chem. *6*, 1720 (1967).
77. *Lang, G., Asakura, T., Yonetani, T.:* Phys. Rev. Letters (to be published).
78. *Moss, T. H., Bearden, A. J., Caughey, W. S.:* J. Chem. Phys. *51*, 2624 (1969).
79. *Koenig, D. F.:* Acta Cryst. *18*, 663 (1965).
80. *Griffith, J. S.:* Biopolymers Symp. *1*, 35 (1964).
81. *Johnson, C. E.:* Phys. Letters *21*, 491 (1966).
82. *Richards, P. L., Eberspaecher, H. I., Caughey, W. S., Feher, G.:* J. Chem. Phys. *47*, 1187 (1967).
83. *Feher, G., Richards, P. L.:* In: Magnetic Resonance in Biological Systems,: p. 141; edited by *Ehrenberg, A., Malmström, B. G., Vänngård, T.* Oxford: Pergamon Press 1967.
84. *Blume, M.:* Phys. Rev. Letters *14*, 96 (1965); *18*, 308 (1967).
85. *Perutz, M. F.:* Nature *194*, 914 (1962).
86. — J. Mol. Biol. *13*, 646 (1965).
87. *Muirhead, H., Perutz, M. F.:* Nature *199*, 633 (1963).
88. *Riggs, A.:* J. Biol. Chem. *236*, 1948 (1961).
89. *Benesch, R., Benesch, R. E.:* J. Biol. Chem. *236*, 405 (1961).
90. *Gibson, Q. H., Roughton, F. J. W.:* Proc. Roy. Soc. (London) *B 146*, 204 (1957).
91. *Wyman, J.:* Advan. Protein Chem. *19*, 233 (1964).
92. *Rossi-Fanelli, A., Antonini, E., Caputo, A.:* Advan. Protein Chem. *19*, 73 (1964).
93. *Antonini, E.:* Physiol. Rev. *45*, 128 (1965).
94. *Ohnishi, S., Boeyens, J. C. A., McConnell, H. M.:* Proc. Natl. Acad. Sci. U.S. *56*, 809 (1966).
95. *Gonser, U.:* J. Phys. Chem. *66*, 564 (1962).
96. — *Grant R. W.:* Appl. Phys. Letters *3*, 189 (1963).
97. *Moss, T. H.:* Ph. D. Thesis, Department of Physics. Ithaca, Cornell University 1965.
98. *Karger, W.:* Ber. Bunsenges. Physik. Chem. *68*, 793 (1964).
99. *Maling, J., Weissbluth, M.:* In: Electronic Aspects of Biochemistry, p. 93; edited by *Pullmann, B,* New York: Academic Press 1964.
100. *Karger, W.:* Z. Naturforsch. *17B*, 137 (1962).
101. *Lang, G., Marshall, W.:* J. Mol. Biol. *18*, 385 (1966).
102. — — Biochem. J. *95*, 56 (1965).
103. — — In: Hemes and Hemoproteins, p. 115; edited by *Chance, B., Estabrook, R. W., Yonetani, T.* New York: Academic Press 1966.
104. *Lang, G., Marshall, W.:* In: Mössbauer Effect Methodology, Vol. 2, p. 127; edited by *Gruverman, I.* New York: Plenum Press 1966.
105. *Bradford, E., Marshall, W.:* Proc. Phys. Soc. (London) *87*, 731 (1966).

106. *Weissbluth, M.:* Struct. Bonding *2,* 1 (1967).
107. *Eicher, H., Trautwein, A.:* J. Chem. Phys. *50,* 2540 (1969).
108. — — J. Chem. Phys. *52,* 932 (1969).
109. *Richards, P. L., Brackett, G. C.:* Proceedings of the Inter-American Symposium on Hemoglobins, Caracas, Venezuela, edited by *Bemsky, G.* 1969.
110. *Brackett, G. C., Richards, P. L.:* Chem. Phys. Letters (to be published).
111. *Grant, R. W., Cape, J. A., Gonser, U., Topol, L. E., Saltman, P.:* Biophys. J. *7,* 651 (1967).
112. *Keilin, D., Hartree, E. F.:* Nature *170,* 161 (1952).
113. *Williams, R. J. P.:* In: Hemes and Hemoproteins, p. 135; edited by *Chance, B., Estabrook, R. W., Yonetani, T.* New York: Academic Press 1966.
114. *McDermott, P., May, L., Orlando, J.:* Biophys. J. *7,* 615 (1967).
115. *Lang, G., Herbert, D., Yonetani, T.:* J. Chem. Phys. *49,* 944 (1968).
116. *Cooke, R., Debrunner, P.:* J. Chem. Phys. *48,* 4532 (1968).
117. *Moss, T. H., Bearden, A. J., Bartsch, R. G., Cusanovich, M. A.:* Biochemistry 7, 1583 (1969).
118. *Ehrenberg, A., Kamen, M. D.:* Biochem. Biophys. Acta *102,* 333 (1965).
119. *Chance, B.:* Arch. Biochem. Biophys. *41,* 404 (1952).
120. *George, P.:* Biochem. J. *54,* 267 (1953).
121. *Theorell, H.:* Enzymologia *10,* 250 (1941).
122. *Yonetani, T.:* J. Biol. Chem. *241,* 2562 (1966).
123. *Theorell, H., Ehrenberg, A.:* Arch. Biochem. Biophys. *41,* 442 (1952).
124. *Morita, Y., Mason, H. S.:* J. Biol. Chem. *210,* 2654 (1965).
125. *Lang, G.:* In: Mössbauer Spectroscopy in Biological Systems, edited by *Debrunner, P., Tsibris, J. C. M., Münck, E.* University of Illinois Bulletin, November, 1969, Urbana.
126. *Arnon, D. I.:* In: Non-Heme Iron Proteins: Role in Energy Conversion, edited by *San Pietro, A.* Yellow Springs, Ohio: Antioch Press 1965.
127. *Hind, G., Olson, J. M.:* Ann. Rev. Plant Physiol. *19,* 249 (1968).
128. *Vernon, L. P., Avron, M.:* Ann. Rev. Biochem. *34,* 269 (1965).
129. *Malkin, R., Rabinowitz, J. C.:* Ann. Rev. Biochem. *36,* 113 (1967).
130. *Mortenson, L. E., Valentine, R. C., Carnahan, J. E.:* Biochem. Biophys. Res. Commun. *7,* 448 (1962).
131. *Tagawa, K., Arnon, D. I.:* Nature *195,* 537 (1962).
132. *Bayer, E., Parr, W., Kazmaier, B.:* Arch. Pharm. *298,* 196 (1965).
133. *Buchanan, B. B., Lovenberg, W., Rabinowitz, J. C.:* Proc. Natl. Acad. Sci. U.S. *49,* 345 (1963). — *Lovenberg, W., Buchanan, B. B., Rabinowitz, J. C.:* J. Biol. Chem. *238,* 3899 (1963).
134. *Malkin, R., Rabinowitz, J. C.:* Biochemistry *5,* 1262 (1966); Biochem. Biophys. Res. Commun. *23,* 822 (1966).
135. *Arnon, D. I.:* Naturwissenschaften *56,* 295 (1969).
136. *Keresztes-Nagy, S., Perini, F., Margoliash, E.:* J. Biol. Chem. *244,* 5955 (1969).
137. *Benson, A. M., Yasunobu, K. T.:* J. Biol. Chem. *244,* 955 (1969).
138. *Sugeno, K., Matsubara, H.:* Biochem. Biophys. Res. Commun. *32,* 951 (1968).
139. *Matsubara, H., Sasaki, R. M.:* J. Biol. Chem. *243,* 1732 (1967).
140. *Mayhew, S. G., Petering, D., Palmer, G., Foust, G. P.:* J. Biol. Chem. *244,* 2830 (1969).
141. *Palmer, G., Brintzinger, H., Estabrook, R. W.:* Biochemistry *6,* 1658 (1967).
142. *Beinert, H., Palmer, G.:* Advan. Enzym. *27,* 105 (1965).
143. — *De Vartanian, D. V., Hemmerich, P., Beeger, C., VanVoorst, J. D. W.:* Biochem. Biophys. Acta *96,* 530 (1965).

144. *VanVoorst, J. D. W., Hemmerich, P.:* In: Magnetic Resonance in Biological Systems, p. 183; edited by *Ehrenberg, A., Malmström, B. G., Vänngård, T.* Oxford: Pergamon Press 1967.
145. *Blumberg, W. E., Peisach, J.:* In: Non-Heme Iron Proteins: Role in Energy Conversion, p. 101; edited by *San Pietro, A.* Yellow Springs, Antioch Press 1965.
146. *Brintzinger, H., Palmer, G., Sands, R. H.:* J. Am. Chem. Soc. **88**, 623 (1966).
147. — — — Proc. Natl. Acad. Sci. U.S. **55**, 397 (1966).
148. *Gibson, J. F., Hall, D. O., Thornley J. H. M., Whatley, F. W.:* Proc. Natl. Acad. Sci. U.S. **56**, 987 (1966).
149. *Johnson, C. E., Bray, R. C., Cammack, R., Hall, D. O.:* Proc. Natl. Acad. Sci. U.S. **63**, 1234 (1969).
150. *Thornley, J. H. M., Gibson, J. F., Whatley, F. R., Hall, D. O.:* Biochem. Biophys. Res. Commun. **24**, 877 (1966).
151. *Orbach, R.:* Proc. Roy. Soc. (London) *A 264*, 458 (1961).
152. *Moss, T. H., Bearden, A. J., Cusanovich, M. A., Bartsch, R. G., San Pietro, A.:* Biochemistry **7**, 1591 (1968).
153. *Dunham, W. R., Bearden, A. J., Sands, R. H., Salmeen, I., Palmer, G., Orme-Johnson, W. H.:* to be published.
154. *Johnson, C. E., Hall, D. O.:* Nature **217**, 446 (1968).
155. *— Elstner, E., Gibson, J. F., Benfield, G., Evans, M. C. W., Hall, D. O.:* Nature **220**, 1291 (1968).
156. *Cooke, R., Tsibris, J. C. M., Debrunner, P. G., Tsai, R., Gunsalus, I. C., Frauenfelder, H.:* Proc. Natl. Acad. Sci. U.S. **59**, 1045 (1968).
157. *Novikov, G. V., Syrtsova, L. A., Likhtenshtein, G. I., Trukhtanov, V. A., Rachek, V. F., Goldanskii, V. I.:* Dokl. Akad. Nauk. SSSR *181*, 1170 (1968).
158. *Wickman, H. H.:* Nuclear and Magnetic Resonance Studies in S-State Ions; Ph. D. Thesis, Lawrence Radiation Laboratory, Berkeley, California, 1965.
159. *Moss, T. H., Petering, D., Palmer, G.:* J. Biol. Chem. **244**, 2275 (1969).
160. *Lewis, J., Mabbs, F. E., Richards, A.:* J. Chem. Soc. (A) 1014 (1967).
161. *Khedekar, A. V., Lewis, J., Mabbs, F. E., Weigold, H.:* J. Chem. Soc. (A) 1561 (1967).
162. *Gerloch, M., Lewis, J., Mabbs, F. E., Richards, A.:* J. Chem. Soc. (A) 112 (1968).
163. *Koenig, E.:* In: Physical Methods in Inorganic Chemistry, edited by *Hill, H. A. O., Day, P.* New York: Wiley/Interscience 1968.
164. *Salmeen, I.:* Ph. D. Thesis, Department of Physics, University of Michigan, Ann Arbor, 1969.
165. *Dunham, W. R.:* Ph. D. Thesis, Department of Chemistry, University of California, San Diego, La Jolla, California, 1970.
166. *Title, R. S.:* Phys. Rev. *131*, 623 (1963).
167. *Oosterhuis, W. T., Lang, G.:* Phys. Rev. *178*, 439 (1969).
168. *Kimura, T.:* Struct. Bonding **5**, 1 (1968).
169. *Yang, C. S., Huennekens, F. M.:* Biochem. Biophys. Res. Commun. **35**, 634 (1969).
170. *Newman, D. J., Postgate, J. R.:* European J. Biochem. **7**, 45 (1968).
171. *Newman, D. J., Ihle, J. N., Dure, L.:* Biochem. Biophys. Res. Commun. **36**, 947 (1969).
172. *Hong, J. S., Rabinowitz, J. C.:* Biochem. Biophys. Res. Commun. **29**, 246 (1967).
173. *Buchanan, B. B., Shanmugan, K. T.:* private communication.
174. *Blomstrom, D. C., Knight, E., Jr., Phillips, W. D., Weiher, J. F.:* Proc. Natl. Acad. Sci. U.S. *51*, 1085 (1964).

175. *Bearden, A. J., Orme-Johnson, W. H.:* unpublished data.
176. *Poe, M., Phillips, W. D., McDonald, C. C., Lovenberg, W.:* Proc. Natl. Acad. Sci. U.S. *65,* 797 (1970).
177. *Orme-Johnson, W. H., Beinert, H.:* Biochem. Biophys. Res. Commun. *36,* 337 797 (1969).
178. *Hong, J. S.:* Ph. D. Thesis, Department of Biochemistry, University of California, Berkeley, California, 1970.
179. *Ghiretti, F.:* In: Oxygenases, pp. 517—553; edited by *Hayaishi, O.* New York: Academic Press 1962.
180. *Klotz, I. M., Keresztes-Nagy, S.:* Biochemistry *2,* 455 (1963).
181. *Okamura, M. Y., Klotz, I. M., Johnson, C. E., Winter, M. R. C., Williams, R. J. P.:* Biochemistry *8,* 1951 (1969).
182. *Klotz, I. M., Klotz, T. A., Fiess, H. A.:* Arch. Biochem. Biophys. *68,* 284 (1957).
183. *Klotz, I. M., Klotz, T. A.:* Science *121,* 477 (1955).
184. *Williams, R. J. P.:* Science *122,* 558 (1955).
185. *Boeri, E., Ghiretti-Magaldi, A.:* Biochem. Biophys. Acta *23,* 489 (1957).
186. *Moss, T. H., Moleski, C., York, J. L.:* J. Biol. Chem. (to be published).
187. *— Bearden, A. J.:* Biochemistry (in press).
188. *Gilchrist, J. L.:* Master's Thesis; Department of Chemistry, University of Texas, Austin, 1965.
189. *Garbett, K., Darnall, D. W., Klotz, I. M., Williams, R. J. P.:* Arch. Biochem. Biophys. *135,* 419 (1969).
190. *Rill, R. L., Klotz, I. M.:* Arch. Biochem. Biophys. *136,* 507 (1970).
191. *McGinnety, J. A., Payne, N. C., Ibers, J. A.:* J. Am. Chem. Soc. *91,* 6301 (1969).
192. *Calligaris, M., Nardin, G., Randaccio, L.:* Chem. Commun. 673 (1969).
193. *Christoph, G. G., Marsh, R. E., Schaeffer, W. P.:* Inorg. Chem. *8,* 291 (1969).
194. *Hall, D. O., Evans, M. C. W.:* Nature *223,* 1342 (1969).
195. *Lang, G.:* Quart. Rev. Biophys. *3,* 1 (1970).
196. *Debrunner, P.:* In: Spectroscopic Approaches to Biomolecular Conformation, edited by *Urry, D. W.* American Medical Association Press 1969.

Received May 19, 1970

Structural Studies of Hemes and Hemoproteins by Nuclear Magnetic Resonance Spectroscopy

Dr. K. Wüthrich

Institut für Molekularbiologie und Biophysik,
Eidgenössische Technische Hochschule, Zürich, Switzerland

Table of Contents

K. Wüthrich

I. Introduction

During the last decade studies of hemoproteins resulted in fundamental contributions to our knowledge of the molecular structures in biological systems. In particular the optical and magnetic properties of the heme groups in these molecules were extensively examined, and thus became a major source of information on structure-function correlations. *Weissbluth (106)* recently discussed the interpretation of data obtained from magnetic suszeptibility measurements, electron paramagnetic resonance (EPR) studies, optical spectroscopy, and Mössbauer spectroscopy. The present review is on nuclear magnetic resonance (NMR) spectroscopy which, with the availability of high resolution spectrometers, has become particularly attractive for studies of the electronic structure of the heme groups in hemoproteins.

Even without all the data on the properties of the heme groups some hemoproteins would be among the best known protein molecules. The amino acid sequences of the polypeptide chains are available for myoglobins, hemoglobins, and cytochromes c from many different species (22). Myoglobin was the first protein of which a detailed molecular structure was obtained by X-ray crystallography (*Kendrew et al. (51—53)*), and today the three-dimensional structures in single crystals are also available of hemoglobin (*Perutz et al. (13, 89—91)*) and cytochrome c (*Dickerson et al. (25, 26)*). These structural data are of great value for the interpretation of the results obtained by other methods. On the other hand, as will be seen throughout this review, NMR spectroscopy can in many instances provide complementary information on the molecular structures in solution.

For a long time hemoprotein research has attracted scientists from different fields, and physicists had a prominent position in many projects. In all a wider variety of physical measurements have been successfully applied to hemoproteins than to any other class of biological molecules, and at present hemoproteins are probably the best known biological molecules from the standpoint of physics. It is therefore not surprising that their investigation is often quoted as an excellent example of research in biophysics, a rapidly expanding area which has grown out of the traditions of the fields of biology, chemistry, medicine, and physics.

In the following sections we shall first present some aspects of structure and function of hemes and hemoproteins, and then proceed to a presentation of some facts on NMR spectroscopy. Sections IV through VI deal with high resolution proton NMR experiments where information was obtained from direct observation of the hemoprotein resonances. This will be follwed by a brief survey of nuclear relaxation enhancement studies, in which the influence of the hemoproteins on the nuclear resonances of the solvent was studied.

II. Some Properties of Hemoproteins

Hemoproteins are involved in many vital processes in living organisms, and a variety of biological roles have been observed for different compounds of this class. Of the molecules discussed below hemoglobin is the oxygen-transporting protein of blood, myoglobin binds and stores oxygen in the muscles, and cytochrome c acts as an electron-transferring oxidation-reduction carrier in the "respiratory chain". Other hemoproteins function as enzymes which control diverse biochemical reactions. Even though the active sites seem to be at or near the heme groups only hemoproteins as entities perform the various biological roles, whereas greatly different chemical reactivities are observed for the isolated heme groups. Therefore much insight into structure — function correlations in hemoproteins can be obtained through detailed studies of the heme-polypeptide interactions in these molecules.

A. Chemistry and Structure

A hemoprotein molecule consists of one or several polypeptide chains and heme groups. The polypeptide chains are strings of amino acid residues linked together by peptide bonds. In the molecular structures determined by X-ray crystallography of protein single crystals the polypeptide chains are uniquely arranged in space. The molecules are fixed in these three-dimensional structures by a multitude of weak bonds, i.e. hydrogen bonds, ionic bonds, and van der Waals interactions, and in some cases also by covalent bonds. In hemoproteins one or several heme groups are located in crevices formed by the three-dimensional arrangement of the polypeptide chains (Fig. 1). They are bound to the latter by covalent (Fig. 2) or coordinative (Fig. 3) bonds, and by weaker interactions.

1. Porphyrin-Iron Complexes

The isolated heme groups in Fig. 2 have molecular weights from 360 to 600, and contain between 12 and 34 protons per molecule. They consist of a porphyrin ring, which is a planar conjugated ring system, and an iron ion which is bound to the four nitrogen atoms of the porphyrin. The

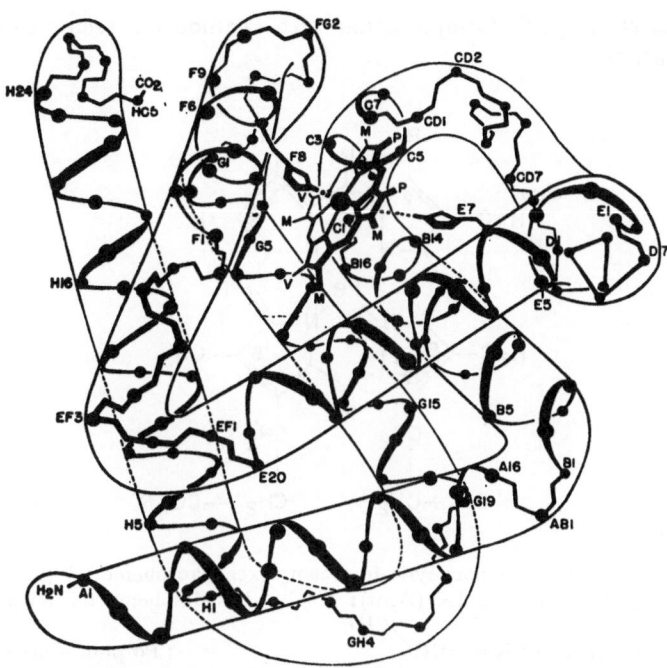

Fig. 1. α-Carbon diagram of myoglobin molecule obtained from 2-Å X-ray analysis in a single crystal. Large dots represent α-carbon positions; labelling is that of *Kendrew et al.* (53). Stretches of α-helix are represented by smooth helix with exaggerated perspective and are given letter-number labels. Nonhelical regions are designated by letter-letter-number symbols, and represented by three-segment zigzag lines between α-carbons. The heme group framework is sketched in forced perspective, with side groups identified by: M = methyl, V = vinyl, P = propionic acid. (Reproduced by permission of *R. E. Dickerson* from ref. (24))

heme iron is usually either in the ferrous or the ferric oxidation state (Table 1). In the presence of molecular oxygen porphyrin-iron(II) complexes in solution are immediately oxidized to the more stable ferric compounds. The properties of solutions of porphyrin-iron(III) complexes depend strongly on the solvent, which is propably due to the tendency of porphyrins to dimerize in solution[1]. Upon addition of KCN to solutions of chloroporphyrin-iron(III) complexes or similar compounds the corresponding cyanide complexes are obtained. In these both axial positions of the heme iron (Fig. 3) are presumably occupied by cyanide ions

---

[1] *W. S. Caughey*, private communication.

(*Hogness et al. (42)*). Complexation with cyanide ion seems to prevent dimerization.

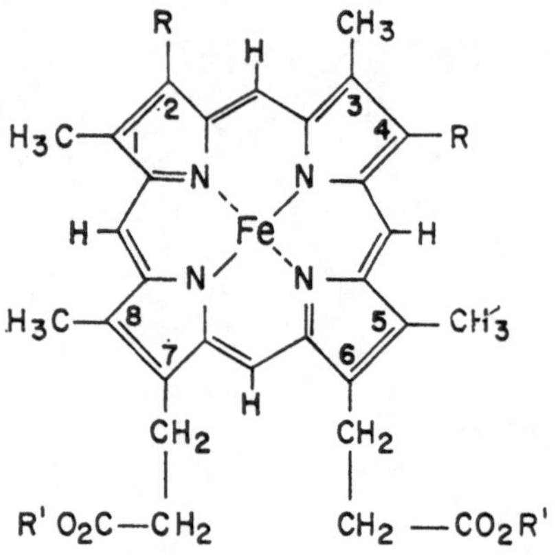

Fig. 2. Structure of some porphyrin-iron complexes. Protoheme IX (Proto): R = —CH=CH₂; Deuteroheme IX (Deut): R = —H; Mesoheme IX (Meso): R = —CH₂—CH₃; Heme c: R = —CH

$$\begin{array}{l} \diagup CH_3 \\ \diagdown S-Polypeptide\ chain \end{array}$$

; Porphin-iron complexes:

All the substituents 1 to 8 are protons. R′ = —H in the heme groups of myoglobin, hemoglobin, and cytochrome c.

2. Myoglobin

Myoglobin has a molecular weight of 18000 and contains ca. 1000 protons per molecule. The complete sequence of the 153 amino acids in the polypeptide chain of the myoglobins from sperm whale (*Edmundson (29)*), horse (*Dautrevaux,* in ref. (22)) porpoise and seal (*Bradshaw and Gurd (15)*) are known, and single crystals of the sperm whale protein were studied extensively by X-ray crystallography (*Kendrew et al. (51—53)*). Myoglobin contains one protoheme IX group (Fig. 2). There are no covalent bonds between protoheme IX and the polypeptide chain, but the histidyl residue at position F-8 is one of the axial ligands of the heme iron (Fig. 1). The function of the sixth coordination site is to reversibly bind and release molecular oxygen (Fig. 3). The heme group and the polypeptide chain of myoglobin can be separated, and then recombined. Myoglobin reconstituted with protoheme IX was found to be identical with the native compound, whereas slightly modified properties were observed when protoheme IX was replaced by deutero-heme IX or mesoheme IX (*Rossi-Fanelli et al. (95)*; *Shulman et al. (98)*).

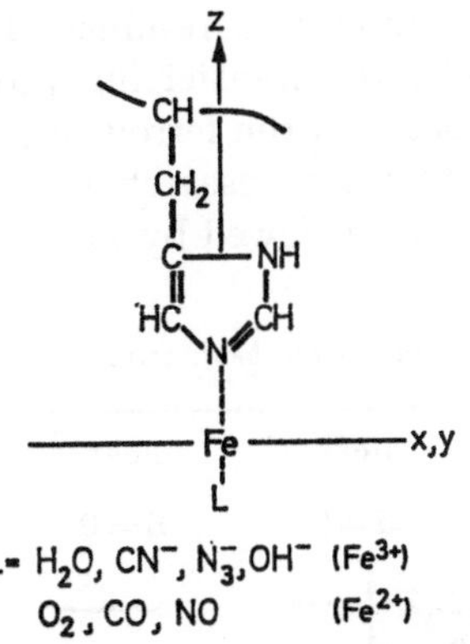

Fig. 3. Axial ligands of the heme iron in hemoproteins. The imidazole ring of a histidyl residue is one of the axial ligands in myoglobin, hemoglobin, and cytochrome c. In native cytochrome c the sixth ligand is a methionyl residue of the polypeptide chain. In partially denatured cytochrome c, and in myoglobin and hemoglobin, a variety of ligands, some of which are shown in the figure, may bind to the sixth coordination site

3. Hemoglobin

Mammalian Hemoglobins have molecular weights of ca. 65000, and consist of four subunits, two each of two types, called α and β. Each subunit contains one polypeptide chain of ca. 150 amino acid residues, and one protoheme IX group. The amino acid sequence is known for a number of hemoglobins from different species (22). The three-dimensional structure of the individual subunits is quite similar to that of the myoglobin molecule (*Perutz et al.* (80, 89—91)), and each of the four heme groups binds one oxygen molecule (Fig. 3). From studies of structure-function correlations in hemoglobin, which were recently reviewed by *Rossi-Fanelli et al.* (95) and by *Antonini* (3, 4), the interactions between the four subunits have long been recognized as an important factor in the biological functions of hemoglobin. Many of the recent studies of subunit interactions involved experiments with separated α- and β-chains, and with mixed state hemoglobin tetramers, where the individual subunits have different electronic configurations.

4. Cytochrome c

Cytochrome c has a molecular weight of 12500, and contains one group of heme c per molecule. The sequence of the 103 to 109 amino acids of the polypeptide chain is known for ca. thirty different species (22). Heme c is covalently linked to the cysteinyl residues at positions 14 and 17 of the polypeptide chain, and the two axial positions of the heme iron are occupied by amino acid residues. The biological function of cytochrome c as an electron-transferring protein is intimately linked with interconversions between the ferrous and ferric oxidation states of the heme iron. The extensive research on structure and function of cytochrome c was recently reviewed by *Margoliash* and *Schejter* (71).

Table 1. *Electron configurations of the heme iron*

oxidation state	Fe^{2+}	Fe^{2+}	Fe^{3+}	Fe^{3+}
spin state	$S = 2$	$S = 0$	$S = 5/2$	$S = 1/2$
electron configuration				
T_{1e}	$\cdots\cdots$	$---$	$1 \cdot 10^{-10}$sec	$2 \cdot 10^{-12}$sec
Examples	Mb^{II}	$Mb^{II}O_2$	$Mb^{III}(H_2O)$	$Mb^{III}CN$
	Hb^{II}	$Hb^{II}O_2$	$Hb^{III}(H_2O)$	$Hb^{III}CN$
		Cyt c^{II}		Cyt c^{III}

T_{1e} = longitudinal electron spin relaxation time.
Mb^{II} = deoxymyoglobin, $Mb^{II}O_2$ = oxymyoglobin, $Mb^{III}(H_2O)$ = ferrimyoglobin, $Mb^{III}CN$ = cyanoferrimyoglobin.
Hb^{II} = deoxyhemoglobin, $Hb^{II}O_2$ = oxyhemoglobin, $Hb^{III}(H_2O)$ = ferrihemoglobin, $Hb^{III}CN$ = cyanoferrihemoglobin.
Cyt c^{II} = reduced cytochrome c or ferrocytochrome c, Cyt c^{III} = oxidized cytochrome c or ferricytochrome c.

B. Electronic Configurations of the Heme Iron

Depending on the oxidation state and the ligand bound at the sixth position of the heme iron (Fig. 3) one of the electronic configurations of Table 1 is commonly observed. The low spin ferrous state (Fe^{2+}, $S = 0$) is diamagnetic, whereas the high spin ferrous (Fe^{2+}, $S = 2$) and ferric (Fe^{3+}, $S = 5/2$) states, and the low spin ferric state (Fe^{3+}, $S = 1/2$) are paramagnetic. The three paramagnetic states have different total electronic spins and electron spin relaxation times, both of which largely influence the NMR spectra.

In the biologically active forms the heme iron of myoglobin and hemoglobin is in the ferrous oxydation state. When no ligand is attached to the sixth coordination site the paramagnetic high spin state is observed, whereas $Mb^{II}O_2$ and $Hb^{II}O_2$ are diamagnetic (*Pauling* and *Coryell* (88)). In vitro myoglobin and hemoglobin can easily be oxidized to the ferric oxidation state. Both oxidized and reduced cytochrome c is in the low spin state (*Theorell (105)*).

C. Oxygen Binding to Myoglobin and Hemoglobin

The oxygenation curve for myoglobin has the shape expected for a bi-molecular reaction. On the other hand the sigmoidal shape of the hemo-globin oxygenation curve (Fig. 4) indicates that oxygen binding to the four subunits is "cooperative". This means that the affinity for oxygen or other ligands of the partially ligated hemoglobins formed through binding of one, two or three ligand molecules is greater than that of deoxyhemoglobin. The free energy of these subunit interactions is ca. 3 Kcal/mol, which is about 10% of the total free energy for complete oxygenation of the hemoglobin tetramer (*Wyman (118)*).

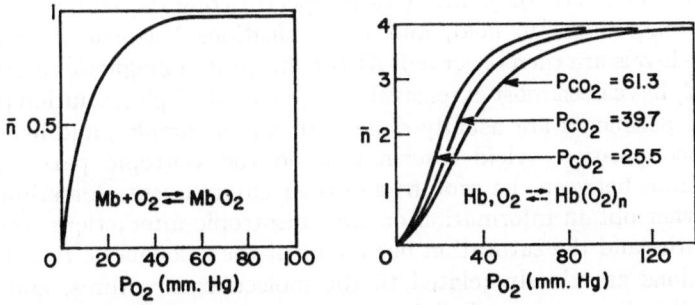

Fig. 4. Oxygenation curves for myoglobin and hemoglobin. The average number of oxygen molecules bound per protein molecule, $\bar{n}$, is plotted vs. the partial pressure of O_2. For hemoglobin the oxygenation curve is given at three different partial pressures of carbon dioxide

Fig. 4 shows further that the oxygen affinity of myoglobin is markedly greater than that of hemoglobin. An interesting feature of the hemoglo-bin oxygenation curve is its dependence on the partial pressure of CO_2 (Fig. 4) and on pH, which is usually referred to as the "Bohr effect" (*12*). From Perutz's X-ray studies (*91*) and chemical investigations by *Kilmartin* and *Rossi-Bernardi* (*54*) the Bohr effect has been related to specific structural features in the hemoglobin molecule. Myoglobin shows no Bohr effect (*95*).

D. Nomenclature

Shorthand notations for the isolated heme groups are given in Fig. 2. For example ProtoCN will stand for cyanoferriprotoporphyrin IX, and we will use DeutCN and MesoCN correspondingly.

Table 1 gives the names and abbreviations used for the hemoproteins. The reconstituted cyanoferrimyoglobins, where protoheme IX was replaced by deuteroheme IX or mesoheme IX, will be referred to as DeutMb^{III}CN and MesoMb^{III}CN. For mixed state hemoglobins the states of the individual subunits will be indicated. For example in Hb($\alpha_2^{II}\beta_2^{III}$CN) the α-chains would be in the deoxy-form, and the β-chains in the cyanoferri-form.

III. Nuclear Magnetic Resonances of Diamagnetic and Paramagnetic Molecules

In nuclear magnetic resonance (NMR) spectroscopy the sample is placed in a strong magnetic field, and the transitions between the nuclear Zeeman levels are then observed. At present proton magnetic resonances (Table 2) have been most extensively investigated. High resolution proton NMR experiments are usually done with liquid samples, and hence the resonance positions yield information on the isotropic parts of the interactions between the protons and their environment. Sometimes one can further obtain information on the anisotropic interactions from the line widths and the saturation behaviour of the resonances. As all these interactions are closely related to the molecular structures, and since essentially all organic and biological compounds contain protons, high resolution proton NMR spectroscopy is widely used for structural studies. NMR spectroscopy has been surveyed in a number of recent textbooks, for example by *Bovey (14)*, and by *Carrington* and *McLachlan (16)*.

Table 2. *Magnetic properties of the proton*

Nuclear Spin Quantum Number	$I = 1/2$; $m_I = \pm 1/2$
Gyromagnetic ratio	$\gamma_H = 2.675 \cdot 10^4$ radians sec^{-1} gauss^{-1}
Resonant frequency at 14 100 gauss =	60 Mc (Megacycles/sec)
23 500 gauss =	100 Mc
51 700 gauss =	220 Mc

A. The Nuclear Magnetic Resonance Experiment

The resonance condition for a nucleus with spin I in a magnetic field H_0 is given by

$$h\nu = \gamma \hbar H_0 \tag{1}$$

where h is Planck's constant, ν is the resonant frequency in cps (cycles/sec), and γ the gyromagnetic ratio of the nucleus. The nuclear resonances are observed either by sweeping the polarizing magnetic field H_0 at constant radio-frequency, or by sweeping the frequency ν at constant field. The common spectrometers operate at field strengths corresponding to proton resonance frequencies of 60 and 100 Mc (Megacycles/sec). For biological applications spectrometers which operate at higher frequencies and use a superconducting solenoid as the source of the polarizing magnetic field (*Nelson* and *Weaver* (*83*)) are also widely used. In most of the studies reported in sections IV through VI a Varian HR-220 spectrometer was employed, which operates in a field-sweep mode at a constant frequency of 220 Mc.

For NMR experiments the compounds can be dissolved in a variety of different solvents, and the sample temperature varied from ca. $-60°$ to 150 °C. Therefore biological systems can be studied under conditions which may be quite similar to the physiological environment of the molecules. Because of the greater effective magnetization at higher polarizing fields the sensitivity of the NMR method increases with increasing operating frequency. The signal-to-noise ratio can be further improved by employing a computer of average transients with observation times of several hours. Biological compounds can be studied in concentrations as low as ca. 0.0005-molar, in certain cases perhaps even less.

B. Diamagnetic Molecules

In organic molecules the different protons are shielded differently from the external magnetic field by their environments, i.e. by the other atoms in the molecule, by the surrounding chemical bonds, and by the solvent. The resulting resonance condition for the ith proton is commonly given by

$$h\nu_i = \gamma_H \, \hbar \, H_0 \, (1 - \sigma_i) \tag{2}$$

where the isotropic screening constant σ_i is a small number of the order 10^{-6}. The observed resonance conditions are usually referred to standard reference compounds, in our experiments either "TMS" (tetramethyl-silane) or "DSS" (2,2-dimethyl-2-silapentane-5-sulfonate). In a field-sweep experiment at constant radio-frequency ν the proton resonances of diamagnetic molecules are observed at lower fields than these standard references within a range of ca. 10 ppm (parts per million) of the applied field. For example, the resonances of aliphatic hydrocarbons are observed between −0.5 and −2.0 ppm, and those of aromatic molecules between −6.0 and −8.5 ppm (Fig. 5). The differences between the resonance positions of the various protons are called "chemical shifts". Since the chemical shifts are proportional to the applied field H_0 (Eq. 2) the spectral resolution is generally increased when higher polarizing fields are applied.

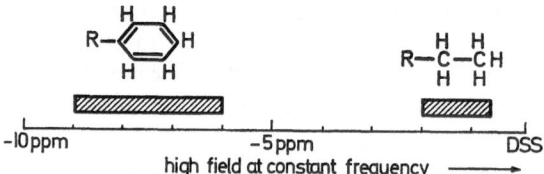

Fig. 5. Proton magnetic resonances of diamagnetic molecules. DSS (2,2-dimethyl-2-silapentane-5-sulfonate) is used as an internal reference. On a HR-220 spectrometer 1 ppm (part per million) corresponds to 220 cps (cycles/sec.)

Each of the resonances which are separated by the chemical shifts may be further split into several lines by spin-spin interactions. These can be described by an isotropic coupling Hamiltonian of the type

$$\mathcal{H} = J \, \vec{I_1} \cdot \vec{I_2} \tag{3}$$

J is called the coupling constant and is measured in cps. $\vec{I_1}$ and $\vec{I_2}$ are the spin quantum numbers of two nuclei in the molecule. For proton-proton coupling the values of $|J|$ are generally between 0 and ca. 20 cps.

The natural line-width of the proton resonances is determined by the nuclear spin relaxation times. Nuclear relaxation in solutions of dia-magnetic molecules comes mainly from intramolecular proton-proton dipolar coupling modulated by the rotational tumbling of the molecules, and will be discussed in more detail in section VII. Suffice it here to say that the line-width increases in general with increasing size of the molecule. For a small organic molecule a line-width at half height of the resonances of less than one cps would be expected, whereas the resonances in proteins with molecular weights of 10 000 to 20 000 may be 20 to 40 cps wide.

C. Ring Current Shifts

The known large diamagnetic anisotropies of aromatic molecules have been explained as being due to a circulation of the π-electrons in the plane of the ring (*Pauling (87)*). This "super-conducting" current produces a local magnetic ring current field H_R which opposes the externally applied field H_0 in the areas above and below the plane of the aromatic molecule, and reinforces it otherwise (Fig. 6).

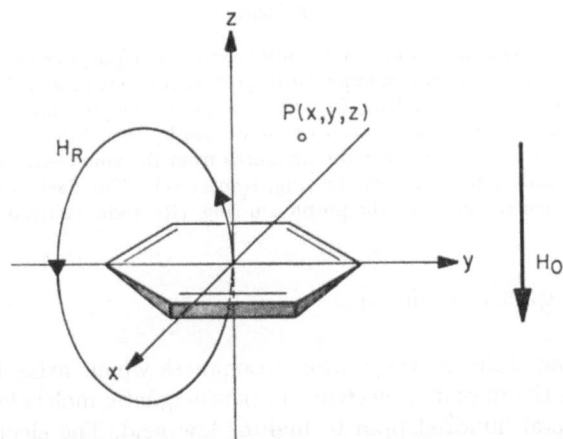

Fig. 6. The local magnetic ring current field H_R of an aromatic molecule. H_0 is the external polarizing field. The field strength at a point P is determined by the position relative to the aromatic ring and by the size of H_R. H_R depends on the total number of π electrons in the aromatic molecule

The shielding zone about a benzene ring has been calculated by *Johnson* and *Bovey* (*48*), and estimates of the ring current fields at positions near a porphyrin ring were compiled by *Shulman et al.* (*99*) from the available experimental data (*Caughey* and *Ibers* (*18*); *Esposito et al.* (*30*)). From these references it can be estimated that the resonance of a proton located near the plane of a phenylalanine ring could be shifted upfield by ca. 2 ppm or less, depending on its position (Fig. 6). A proton near the plane of a heme group might be shifted upfield by as much as 5 ppm, perhaps even more (Fig. 7).

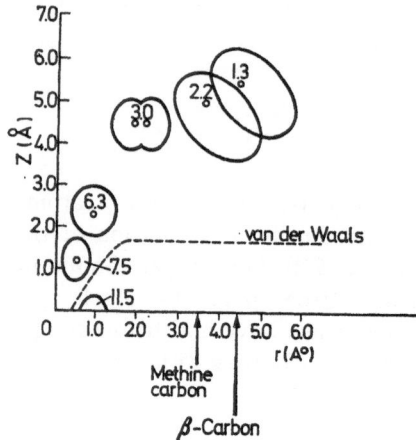

Fig. 7. A plot of the upfield ring current shifts measured in porphyrins, and in phtalo-cyanine model complexes where aliphatic hydrocarbons were attached to the central metal ion (*Caughey* and *Ibers* (*18*); *Esposito et al.* (*30*)). The positions of the observed protons were derived within the accuracy indicated by the ellipses from the known molecular structures. Z and r are the distances from the ring center perpendicular to the plane, and in the plane of the ring, respectively. The dashed line indicates the van der Waal thickness of the porphyrin ring. (Reproduced from ref. (*99*))

D. Paramagnetic Molecules

The hyperfine shifts of the proton resonances which arise from inter-actions with the unpaired electrons in paramagnetic molecules can be as large as several hundred ppm to high or low field. The electron-proton interactions may also broaden the resonances, and the proton NMR spectra yield information on paramagnetic molecules only if the hyperfine shifts are large compared to the line widths. This is generally the case when the electronic spin relaxation times are very short. As an approxi-mate rule hyperfine shifts can be observed in the proton NMR spectra of solutions of paramagnetic metal complexes if the electron paramagnetic resonance signal cannot be detected at room temperature because of excessive line-broadening.

The observed hyperfine shifts could come from contact coupling or "pseudocontact" interactions between the electrons and the protons. Contact shifts arise when a finite amount of unpaired electron density is transferred to the observed protons. The contact shifts of the proton resonances for isotropic systems are given by *Bloembergen's* (*9*) expression

$$\Delta\nu_{cI} = - A_i \frac{|\gamma_e|}{|\gamma_H|} S (S + 1) \frac{\nu}{3 kT} \qquad (4)$$

where A_i is the contact interaction constant for the ith proton, γ_e and γ_H are the gyromagnetic ratios for the electron and the proton, S is the total electronic spin, v the resonance frequency for the proton, k the Boltzmann constant, and T the absolute temperature. Eq. (4) shows that Δv_c is proportional to the reciprocal of temperature. For systems with an anisotropic g-tensor a modified form of Eq. (4) should be employed. This will be further discussed in section IV, together with the pseudocontact interactions.

Scalar electron-proton coupling modulated by the electronic spin flip, and dipole-dipole coupling modulated by the combination of electronic spin flip and rotational tumbling of the molecules contribute to the transverse proton nuclear spin relaxation, and hence to broadening of the resonances. The resulting line broadening in cps at half height of the ith proton resonance, δv_i, may be described by (*Solomon (102)*; *Abragam (1)*).

$$\delta v_i = \frac{1}{\pi T_{2i}^p} = \frac{1}{\pi T_{2i}^{\text{scalar}}} + \frac{1}{\pi T_{2i}^{\text{dipole}}}$$

$$= \frac{1}{3\pi} S(S+1) \left\{ \frac{A_i^2}{\hbar^2} T_{1e} + \gamma_I^2 \gamma_S^2 \hbar^2 \frac{1}{5 r_i^6} \left[7 \tau_c + \frac{13 \tau_c}{1 + \omega_S^2 \tau_c^2} \right] \right\} \tag{5}$$

The transverse proton spin relaxation times T_2^{scalar}, T_2^{dipole}, and T_2^p describe the rates of the nuclear spin relaxation induced respectively by scalar coupling, dipolar coupling, and the sum of the electron-proton interactions. S is the total electronic spin, A_i the scalar coupling constant for the ith proton, $\hbar$ is Planck's constant, T_{1e} the longitudinal electronic relaxation time, γ_I and γ_S are the gyromagnetic ratios for the proton and the electron, and r_i is the electron-proton distance. τ_c is the correlation time for dipole-dipole coupling, with $1/\tau_c = 1/T_{1e} + 1/\tau_r$, where τ_r is the correlation time for rotational tumbling of the molecule, and ω_S is the resonance frequency of the electron in radians/sec. In writing Eq. (5) it was assumed that $\omega_I \tau_c \ll 1$, which is valid for the systems discussed in this paper. Eq. (5) shows that the proton resonance line width may be essentially unaffected by the electronic spin if T_{1e} is sufficiently short.

E. Hemoproteins

The proton resonances of the common amino acids, which were recently surveyed by *McDonald* and *Phillips (78)*, and by *Roberts* and *Jardetzky* *(94)*, are all in the spectral region from DSS to −10 ppm. The spectrum of a protein is also mostly in this region, even though the resonances of the individual amino acid residues are generally somewhat shifted by the interactions with the neighbouring groups in the molecule (*Kowalsky* (*61*); *McDonald* and *Phillips* (*77*); *Sternlicht* and *Wilson* (*103*)). Because of the large size of protein molecules the individual resonances are rather broad, and the resonances of the many hundred to several thousand protons overlap therefore strongly. However, some resonances of protons located near aromatic amino acid side chains or near the heme groups in hemoproteins can be shifted to unusual positions. For example the methyl resonance of an aliphatic amino acid may be resolved at a position upfield from DSS if it experiences a ring current shift of more than ca. 1.5 ppm, and a corresponding ring current shift of a proton resonance originally located at lower fields may in certain favorable cases also yield a resolved line (Fig. 8). Once the resolved ring current shifted lines are identified they can yield information on the three-dimensional molecular structure in solution, since the size of $\Delta\nu_r$ depends very sensitively on the relative positions of the observed protons and the aromatic rings (Figs. 6 and 7).

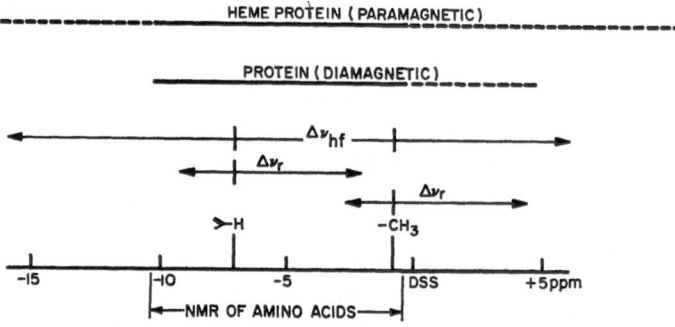

Fig. 8. Scheme of the proton NMR spectra of diamagnetic and paramagnetic hemoproteins. In the spectral region from DSS to −10 ppm, where all the resonances of the common amino acids are observed, we have the strongly overlapping resonances of the polypeptide chains. In diamagnetic molecules some resolved lines may be between DSS and +5 ppm because of local ring current fields. In paramagnetic compounds we may have additional lines at high and low fields which were shifted by the hyperfine interactions between protons and electrons

The NMR spectra of paramagnetic hemoproteins contain resolved hyperfine-shifted lines in addition to the ring-current shifted resonances (*Kowalsky* (*62*); *Kurland et al.* (*64*); *Wüthrich et al.* (*112*)). Between DSS and ca. +5 ppm the spectra can then contain resolved ring current-shifted and hyperfine-shifted resonances (Fig. 8). These can in general be distinguished by their different temperature dependences. The hyperfine-shifted resonances were assigned to the protons of the heme group and the axial ligands of the heme iron (Figs. 2 and 3). Therefore these resonances can yield information on the electronic structure of the heme groups in the hemoproteins.

From sections IIIA through E it would appear that at present high resolution proton NMR spectroscopy cannot generally be applied to very large biological molecules. Increasing size of the molecules increases the difficulties to obtain sufficiently high molar concentrations, and decreases the spectral resolution because of the increase in both the number of protons per molecule, and the resonance line widths. However, resolved hyperfine-shifted resonances may be observed for rather large molecules. For example the spectra of paramagnetic hemoglobins contain a considerable number of resolved lines despite the high molecular weight of 65000 (*Kurland et al.* (*64*); *Wüthrich et al.* (*113*)).

IV. Low Spin Ferric Hemes and Hemoproteins

The hyperfine shifts in the proton NMR spectra of paramagnetic hemes and hemoproteins are closely related to the electronic structures of these molecules. At present the most extensive NMR studies of the electronic spin distribution in the heme groups have been done with low spin ferric compounds, which will be discussed in this section. Procedures similar to those described here would apply to the analysis of the NMR spectra of hemoproteins in the other paramagnetic states (Table 1).

The hyperfine shifts of the proton resonances are obtained as the differences between the observed positions of corresponding resonances in the paramagnetic and diamagnetic compounds. For the present experiments zinc(II)-complexes were chosen as a diamagnetic reference (Fig. 9)[2]. The unusual low-field positions of the proton resonances in the diamagnetic porphyrins arise because of the large ring current field of the porphyrin ring (*Becker et al.* (*5, 6*); *Caughey et al.* (*17, 18*)). In the porphin-zinc(II) complex the four mesoprotons and the protons at positions 1 to 8 are equivalent, whereas the fourfold symmetry is slightly perturbed in the other compounds where different groups are attached to positions 1 to 8. The resulting chemical shifts among the four ring methyl resonances (a) and mesoprotons (b), as well as the dependences of the resonance positions on the groups at positions 2 and 4, the tem-

---

[2] We would like to thank Dr. *W. S. Caughey* for providing the porphyrin-zinc(II) complexes.

perature, and the solvent, are small compared to the hyperfine shifts in the paramagnetic compounds. Therefore the following reference positions for diamagnetic porphyrins were used throughout: ring methyls −3.6 ppm from DSS, mesoprotons −10.5 ppm, and 2,4-protons in deuteroporphyrin IX −9.4 ppm.

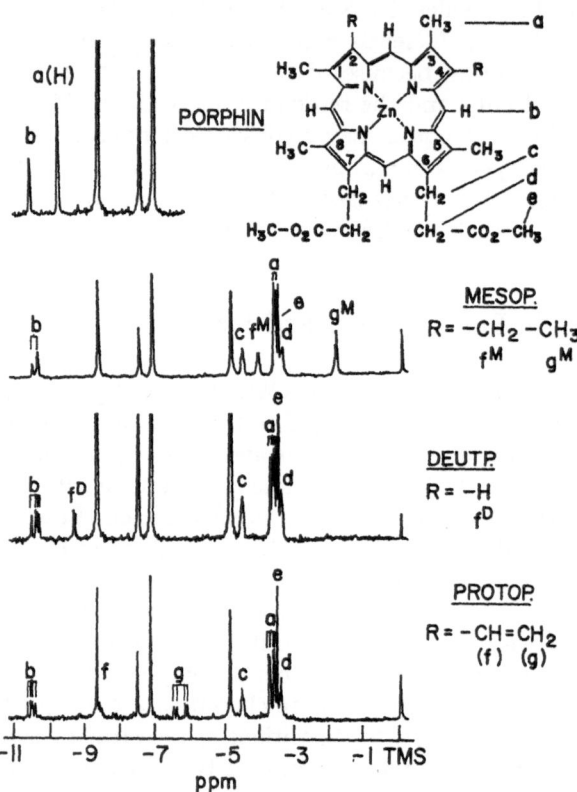

Fig. 9. Proton NMR spectra at 220 Mc of solutions in d_5-pyridine of the Zn(II)-complexes with porphin, and the dimethylesters of mesoporphyrin IX, deuteroporphyrin IX, and protoporphyrin IX. The resonance assignments were based on the relative resonance intensities and the observed fine-structure from spin-spin coupling; they agree with previously published data by *Caughey* and *Koski* (17): a = ring methyls (for porphin: protons at positions 1 to 8), b = mesoprotons, c and d = methylene protons of the propionates, e = methylesters, f and g = resonances of the substituents at positions 2 and 4. Three strong resonances between −7 and −9 ppm come from d_5-pyridine, the line at ca. −4.9 ppm from HDO. $T = 25\,°C$

A. Proton Magnetic Resonance Spectra

In the following the proton NMR spectra of some low spin ferric compounds are presented, and the assignment of the resonances to specific protons in the molecules will be discussed.

1. Cyanoporphyrin-Iron(III) Complexes

In Fig. 10 the proton NMR spectrum of cyanoferriporphin is compared to that of a diamagnetic porphin complex. In both spectra the protons

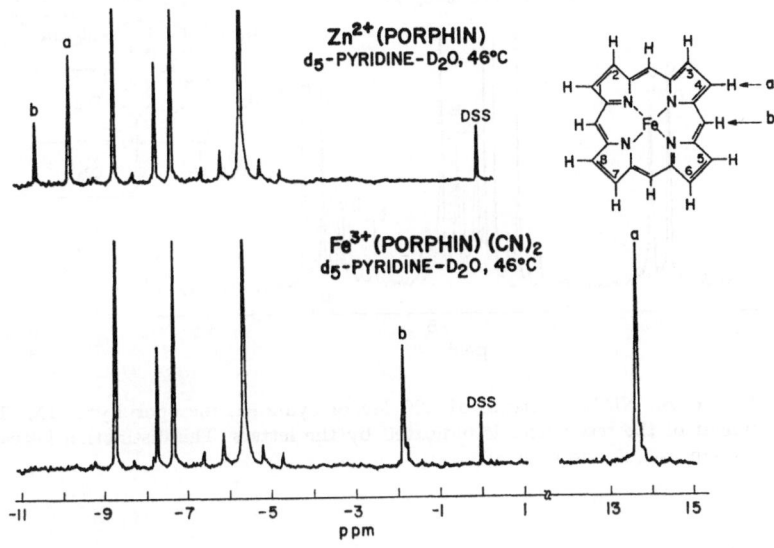

Fig. 10. Proton NMR spectra at 220 Mc of cyanoferriporphin and porphin zinc(II). The lines between —4.5 and —9.5 ppm correspond to the solvent resonances and their spinning side-bands

at positions 1 to 8 (a), and the four mesoprotons (b) are equivalent, as one might have anticipated from the D_{4h} symmetry of the molecular geometry. The resonance b is shifted ca. 9 ppm, the resonance a ca. 24 ppm upfield by the hyperfine interactions with the unpaired electron in cyanoferriporphin. On the other hand the line width of the resonances is almost unaffected by the electronic spin.

Figs. 11 to 13 show the NMR spectra of cyanoferrimesoporphyrin IX, cyanoferrideuteroporphyrin IX dimethylester, and cyanoferriprotoporphyrin IX. The assignments of the resonances were based on their relative intensities, and on comparison of the different compounds (*114*). At ambient temperature the resonances of the ring methyls (a) are shifted ca. 10 to 15 ppm downfield by the hyperfine coupling with the electronic spin, whereas the mesoprotons are shifted 8 to 12 ppm upfield. Further comparison with the diamagnetic porphyrins (Fig. 9) shows that the differences between the resonance positions of the four ring methyls and

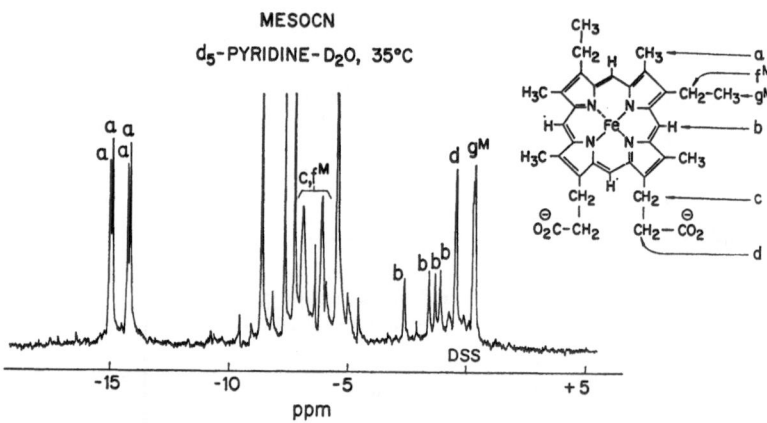

Fig. 11. Proton NMR spectrum at 220 Mc of cyanoferrimesoporphyrin IX. The assignment of the resonances is indicated by the letters. The distinction between the resonances c, d, and f^M is arbitrary

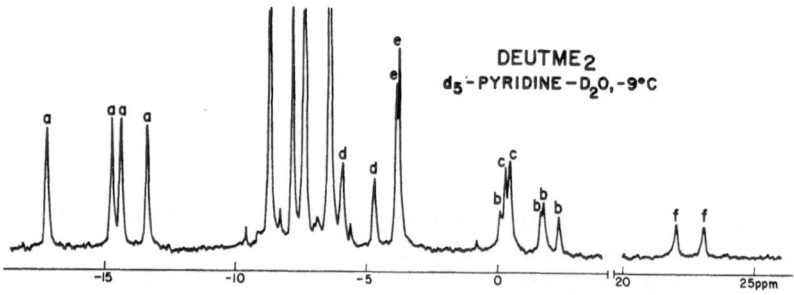

Fig. 12. Proton NMR spectrum at 220 Mc of cyanoferrideuteroporphyrin IX dimethylester. The letters refer to the resonance assignments of Fig. 9. The distinction between the resonances c and d is arbitrary (Reproduced from ref. (*114*))

the four mesoprotons are much larger in the iron(III)-complexes, and hence have to come essentially entirely from different hyperfine interactions with the individual ring methyls and mesoprotons.

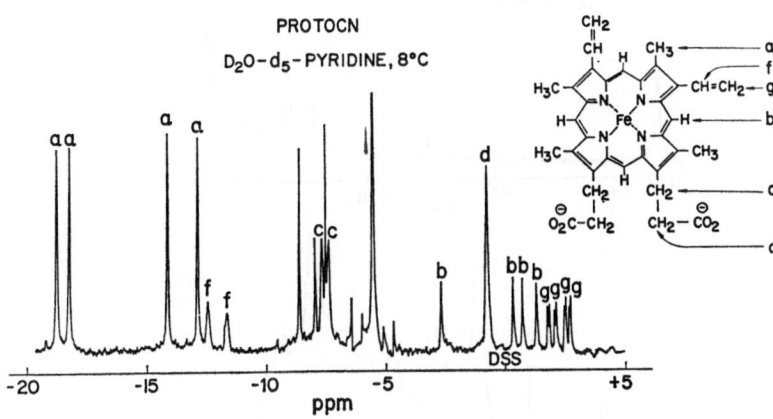

Fig. 13. Proton NMR spectrum at 220 Mc of cyanoferriprotoporphyrin IX. The distinction between resonances *c* and *d* is arbitrary. In the resonances of the 2,4-vinyl groups (*f* and *g*) the fine-structure from proton-proton spin-spin coupling is partially resolved

2. Cyanoferrimyoglobin

The basic features of the cyanoferrimyoglobin spectrum (Fig. 14) are readily understood from the discussion in the preceding sections. In the spectral region from −0.5 to −9 ppm we have the strongly overlapping resonances of almost all the ca. 1000 protons of the polypeptide chain. Outside of this range are a number of resolved resonances which correspond in intensity to one, two or three protons. These lines could have been shifted to their unusual positions by the ring current fields of the aromatic rings, or by hyperfine interactions with the unpaired electron (Fig. 8). Since no protons in a hemoprotein could conceivably experience larger downfield ring current shifts than those of the mesoprotons in diamagnetic porphyrins (Figs. 6 and 9) all the resonances below −10.5 ppm have to be shifted by hyperfine coupling. Accordingly all these lines are temperature-dependent (Eq. 4). On the other hand Figs. 15 and 16 show that the spectral region from −0.5 to 5 ppm of sperm whale cyanoferrimyoglobin contains also some temperature-independent lines. These were assigned to ring current-shifted resonances, which one would

73

expect to be temperature-independent in the absence of conformational changes with temperature (112).

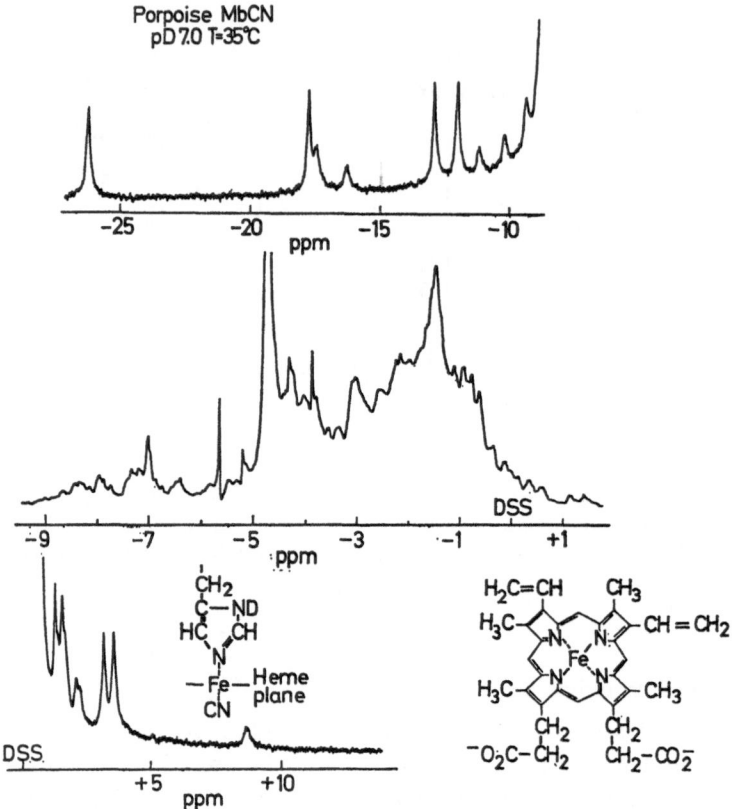

Fig. 14. Proton NMR spectrum at 220 Mc of cyanoferrimyoglobin. Different scales were used for the different spectral regions. Between DSS and −10 ppm there are ca. 1000 protons of the protein, whereas the intensities of the resolved lines in the regions −10 to −25 ppm, and DSS to +10 ppm correspond to one to three protons. The sharp lines between −3.8 and 6 ppm are the resonance of HDO and its first and second spinning side bands. Protoheme IX and the axial ligands of the heme iron in Mb[III]CN are shown at the bottom. (Reproduced from ref. (115))

Only the proton resonances of the heme group and the imidazole ring of the axial histidine (Fig. 14) would be expected to experience sizeable contact shifts, and pseudocontact shifts are probably also largest for these protons. The total intensity of the resolved temperature-

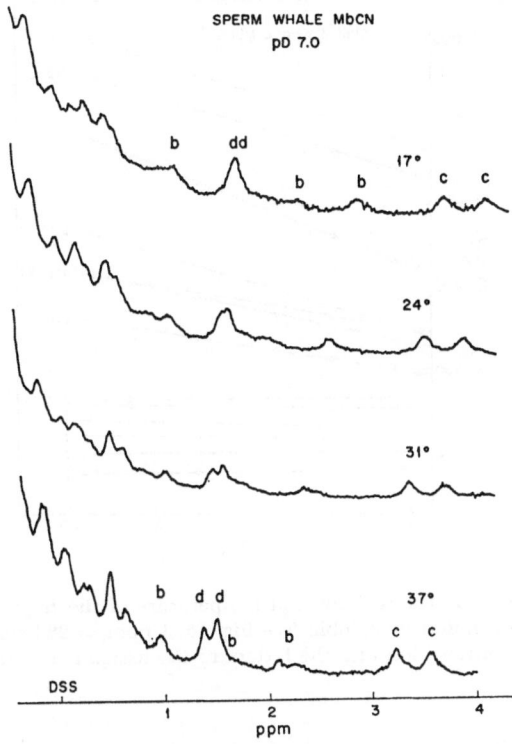

Fig. 15. Dependence on temperature of the spectral region from −0.5 to +5 ppm of sperm whale cyanoferrimyoglobin. The letters refer to the resonance assignments in Fig. 17

dependent resonances was found to correspond almost exactly to the number of protons on the ligands bound directly to the heme iron. Assignments of the hyperfine-shifted resonances to specific protons of the heme group were based on the relative intensities of the resonances, and on comparison with reconstituted DeutMb^{III}CN (*Shulman et al. (98)*). Comparison of the data in Fig. 17 with the resonance assignments in ProtoCN and DeutCN (Figs. 12 and 13) shows that the resonances of the ring methyls, the 2,4-protons, and the mesoprotons are in similar spectral regions, even though the size of the hyperfine shifts differs quite drastically between the isolated hemes and the myoglobins. On the other hand it appears that all the resonances of the propionates and the vinyl groups are not in similar spectral regions for ProtoCN and Mb^{III}CN. Probably this has to be related to the specific sterical orientations of these rather bulky heme-substituents in the myoglobin molecule (*51—53*).

75

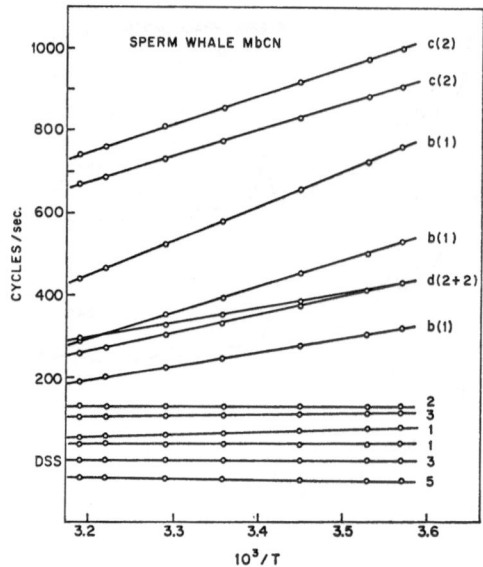

Fig. 16. Dependence on the reciprocal of temperature of the high field resonances of sperm whale cyanoferrimyoglobin (see Fig. 15, 1 ppm = 220 cps). The numbers correspond to the intensities, and the letters to the assignments of the resonances in Fig. 17

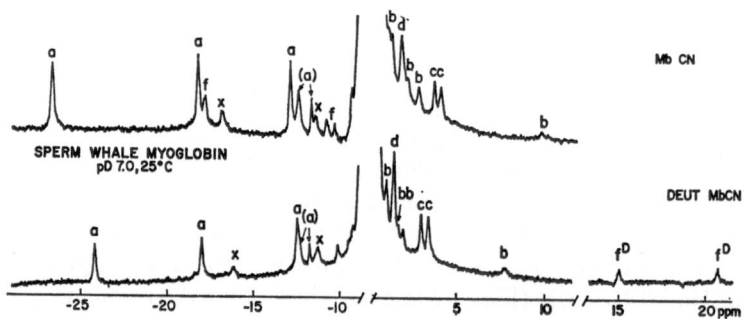

Fig. 17. Proton NMR spectra of native Mb^{III}CN and reconstituted DeutMb^{III}CN (Fig. 2). The resonance assignments are: (a) ring methyls, (b) mesoprotons, (c, d) methylene protons of the propionate groups, (f) —CH protons of the vinyl groups (f^D) 2,4-protons in DeutMb^{III}CN, (x) tentatively assigned to the imidazole protons of the axial histidine. The f-resonances were assigned from their absence in Deut-Mb^{III}CN, the other resonances from their intensities and their presence in both spectra. (Reproduced from ref. (98))

3. Cyanoferrihemoglobin

Because of the higher molecular weight of hemoglobin the resonances in the diamagnetic region of the spectrum overlap even more strongly than in myoglobin (Fig. 18). By similar arguments to those in the preceding section on myoglobin the hyperfine-shifted resonances at high and low fields were assigned to the heme groups. The assignment to specific protons of the heme groups is even more complicated, since there are four heme groups per molecule, two each of the α- and β-subunits. Studies of the isolated subunits showed that the hyperfine-shifted resonances of the different subunits are slightly different. At present it appears most likely that in the spectrum of Fig. 18 two ring methyl resonances of each heme group might be at −21 to −22 ppm and −15 to −16 ppm, whereas the other two ring methyls per heme might not be resolved[3]).

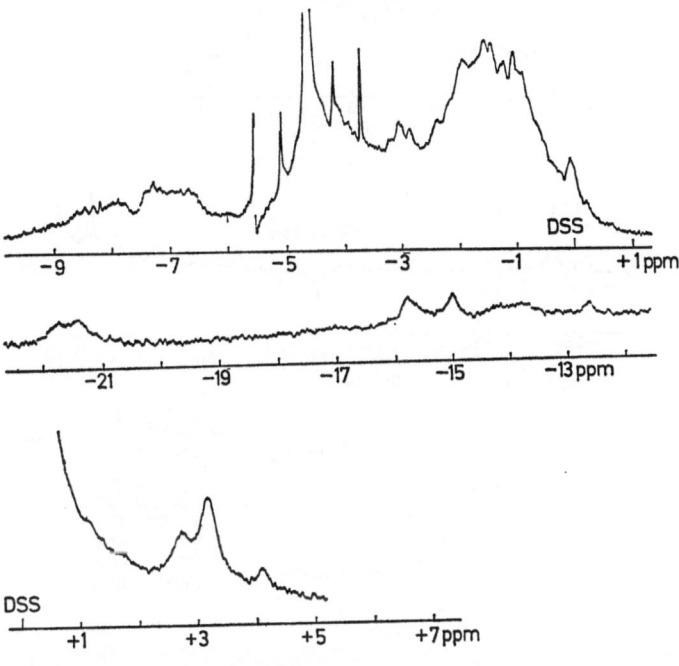

Fig. 18. Proton NMR spectrum at 220 Mc of human cyanoferrihemoglobin at 36 °C. Different scales were used for the different spectral regions. This spectrum was recorded in a single sweep. The signal-to-noise ratio could be further improved by employing a computer of average transients

---

[3]) *S. Ogawa* and *R. G. Shulman*, private communication.

4. Ferricytochrome c

In the spectrum of ferricytochrome c (Fig. 19) the resonances between −0.5 and −9 ppm correspond to ca. 650 protons of the polypeptide chain. Essentially all the proton resonances of heme c and the axial ligands were found to be shifted by hyperfine interactions with the unpaired electron (*Kowalsky (62)*; *Wüthrich (110)*), and some ring current-shifted resonances were also outside of the spectral region from DSS to −10 ppm. From a temperature study similar to that shown for MbCN (Figs. 15 and 16) the temperature dependent hyperfine-shifted resonances in Fig. 20 were identified (*110*). A tentative assignement of some of these lines

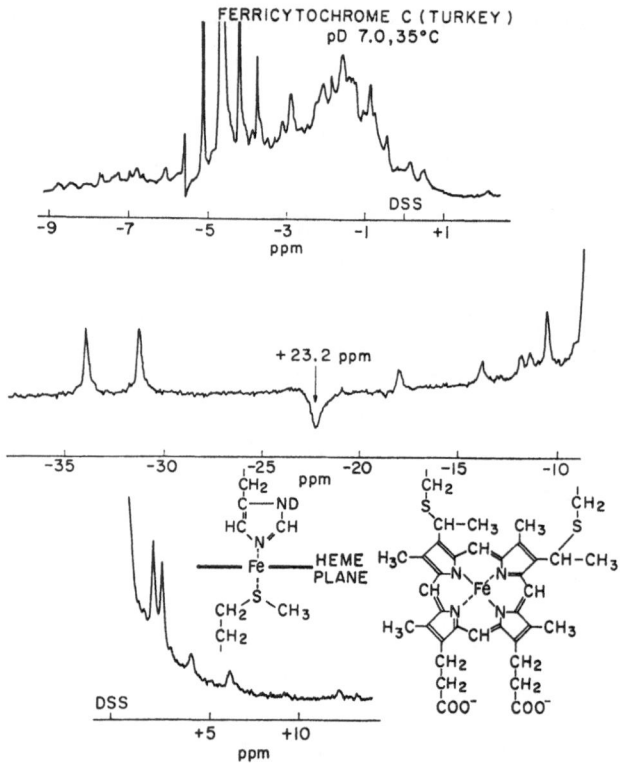

Fig. 19. Proton NMR spectrum at 220 Mc of ferricytochrome c. Different scales were used for the different spectral regions. The high field line at +23.2 ppm was observed as an inversed resonance of the center-band of the spectrum (the HR-220 operates with a 10^4 cps field modulation, and one usually observes the first upfield side-band. With the occurrence of large hyperfine shifts the center band and the side bands sometimes overlap). Heme c and the axial ligands of the heme iron are shown at the bottom. (Reproduced from ref. (*111*))

was based on the relative resonance intensities, and on symmetry considerations. Since only one resonance of intensity corresponding to more than one proton is observed at high fields from $+3$ ppm, at least five of the six methyl groups of heme c (Fig. 19) have to be between -35 and $+3$ ppm. From the symmetry of the electronic wave functions of the porphyrin ring (*Longuet-Higgins et al.* (*68*)) it appears then extremely unlikely that any of the methyl groups of heme c could be shifted to $+23.2$ ppm, and hence this resonance was assigned to the axial methionine. Furthermore the two resonances at -34 and -31.4 ppm were assigned to two ring methyl groups, because otherwise one would have to make the unlikely assumption that the methyl groups in the β-positions of the 2,4-substituents interact much more strongly with the unpaired electron than any of the ring methyls.

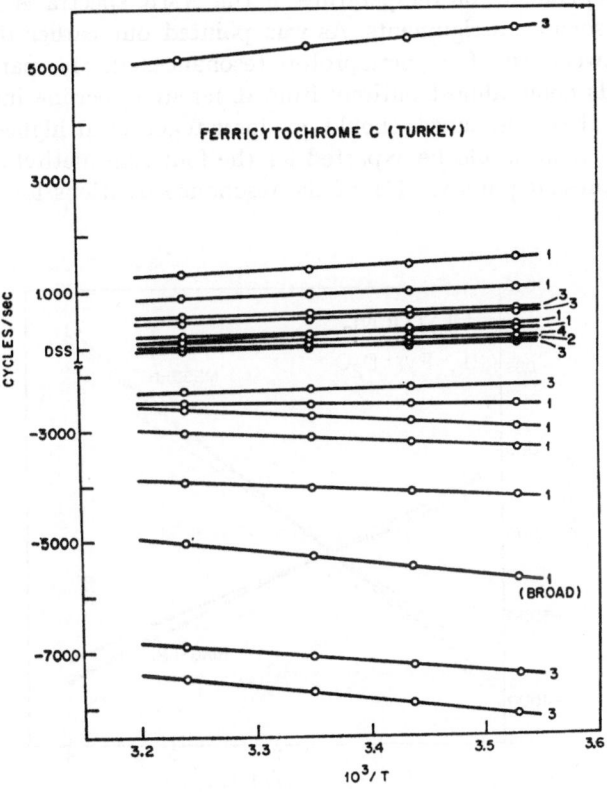

Fig. 20. Dependence on temperature of the resolved hyperfine-shifted resonances of ferricytochrome c. The numbers on the right give the relative intensities of the lines

5. Dependence on Temperature

From Figs. 16 and 20 it would appear that within experimental accuracy the positions of the hyperfine-shifted resonances depend linearly on the reciprocal of temperature, as expected from Eq. (4). Eq. (4), and corresponding expressions for pseudocontact shifts, would further imply that the plots of the resonance positions vs. 1/T should go to the corresponding resonances in the diamagnetic compounds (Fig. 9) when extrapolated to very high temperatures. Because the temperature range covered by the experiments was rather limited such extrapolations would not be expected to be very accurate. Yet it seems that the deviations from the expected behaviour (Fig. 21—23) are quite definitely beyond the limits of experimental error. This appears to indicate that in the temperature range covered by the NMR experiments the scalar coupling constant A in Eq. (4) is probably also dependent on temperature.

The dependence on temperature of the NMR spectra is also useful for the resonance assignments. As was pointed out earlier the relative shifts between the four mesoproton resonances in the paramagnetic compounds come almost entirely from different hyperfine interactions. Therefore these resonances should go closer together at higher temperatures. The same would be expected for the four ring methyl lines. Figs. 21 to 23 present plots vs. 1/T of the resonance positions for meso- and

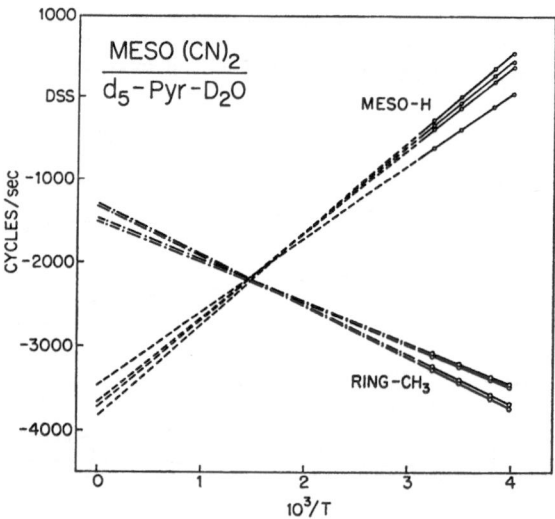

Fig. 21. Dependence on the reciprocal of temperature of the resonance positions for the mesoprotons and the ring methyls in cyanoferrimesoporphyrin IX. The solid lines indicate the temperature range covered by the experiments

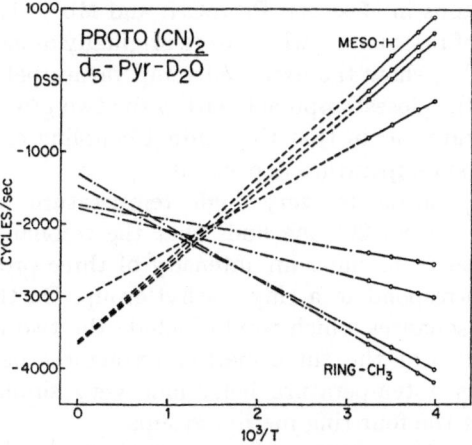

Fig. 22. Dependence on the reciprocal of temperature of the resonance positions for the mesoprotons and the ring methyls of cyanoferriprotoporphyrin IX. The solid lines indicate the temperature range covered by the experiments

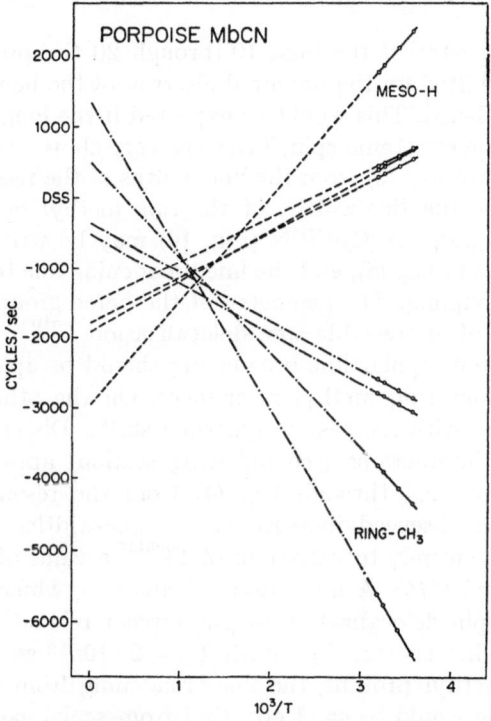

Fig. 23. Dependence on the reciprocal of temperature of the resonance positions for the mesoprotons and the ring methyls in cyanoferrimyoglobin. The solid lines indicate the temperature range covered by the experiments

ring methyl protons in MesoCN, ProtoCN, and Mb^{III}CN. It is seen that the two groups of resonances, which were originally assigned on the basis of their intensities, show the expected temperature behaviour. The observations that the closest approach within the two groups of lines is not at $1/T \approx 0$ indicates again that the contact coupling constant A in Eq. (4) might itself be temperature-dependent.

From extrapolation to very high temperature of the plot for ferricytochrome c (Fig. 20) one finds that the resonance at ca. -10.5 ppm, which appears to have an intensity of three protons, does most probably not correspond to a ring methyl group. On the other hand a group of four resonances which would include the two lines at ca. -33 ppm, and any two of the three methyl resonances between DSS and $+3$ ppm, shows a temperature behaviour very similar to what one might expect for the four ring methyl groups.

B. Electronic Spin Relaxation Times

In the NMR spectra of the Figs. 10 through 20 the proton resonances were sizeably shifted by the unpaired electron of the heme iron, but not markedly broadened. This would be expected if the longitudinal relaxation time for the electronic spin, T_{1e}, were very short. Estimates for T_{1e} were derived with Eq. (5) from the line-widths of the resonances.

For example the line-widths of the ring methyl resonances of the heme group in porpoise Mb^{III}CN (Fig. 14) may be written as the sum of the two terms in Eq. (5), and the line broadening due to proton-proton dipole-dipole coupling. The geometry of the heme group indicates that in the absence of appreciable spin delocalization T_2^{dipole} in Eq. (5), and the proton-proton dipolar line-broadening should be approximately the same for the four ring methyl resonances. On the other hand T_2^{scalar} becomes shorter with increasing hyperfine shifts. Disregarding possible pseudocontact interactions (see following section) approximate values for A_i were obtained through Eq. (4) from the resonance positions. Attributing the observed increase of the line-widths with increasing hyperfine shifts entirely to variations of T_2^{scalar} a value of $T_{1e} \approx 2 \cdot 10^{-12}$ sec was obtained (116). A more detailed analysis, which would include the effects of spin delocalization on the dipolar relaxation, might yield an even somewhat shorter T_{1e}. With $T_{1e} = 2 \cdot 10^{-12}$ sec and $r_i = 6.4$ Å for the ring methyl protons, the line-broadening from electron-proton dipolar coupling would be ca. 1 cps, that from scalar coupling $\leqslant 5$ cps. The observed line-widths of ca. 25 cps then seem to come mainly from proton-proton dipolar coupling. Since the spectral resolution is thus essentially unaffected by the electronic spin the extremely short electron spin relaxation times make low spin ferric hemes and hemoproteins particularly attractive for NMR studies.

C. Kramers Doublets

Theoeretical models proposed by *Griffith* (*36, 38*) and *Kotani* (*59, 60*) are now generally employed for the interpretation of the spectral and magnetic properties of hemoproteins. In his recent review *Weissbluth* (*106*) gave a detailed account of the calculations involved in these models. The present discussion is limited to a brief outline of the treatment of low spin ferric heme compounds, and the presentation of some results which will be useful for the analysis of the NMR data.

In ferric iron the electron configuration outside the closed shells is $(3d)^5$. $(3d)$ stands for the five d-orbitals d_{z^2}, $d_{x^2-y^2}$, d_{xy}, d_{xz}, and d_{yz}, where the z-axis would be perpendicular to, and the x- and y-axes in the heme plane (Fig. 3). Under the influence of an octahedral ligand field the d-orbitals are split into two groups of symmetries e_g and t_{2g}. In low spin ferric compounds the energy separation between the e_g-orbitals, d_{z^2} and $d_{x^2-y^2}$, and the three t_{2g}-orbitals is large compared with spin-pairing energies. Therefore the electronic configuration is $(t_{2g})^5$. Since occupation by five electrons leaves a single hole in the t_{2g}-manyfold, the system behaves precisely as if it contained a single electron. The complicated $(3d)^5$ system can be replaced by the relatively simple $(3d)^1$ "hole" system; for example an electron configuration such as $|(3d_{xy})^2(3d_{xz})^2 (3d_{yz})^-\rangle$ would be replaced by $|(3d_{yz})^+\rangle$. The $(3d)^1$ hole is subject to Kramers theorem which states that when a system is composed of an odd number of electrons it is not possible for electric fields to remove degeneracies completely; at least two-fold degeneracy must remain (*37*).

A detailed description of the electronic states which result from the $(3d)^+$ configuration in azidoferrihemoglobin and azidoferrimyoglobin at low temperatures was derived by *Gibson* and *Ingram* (*35*) and *Griffith* (*36*) from single crystal electron paramagnetic resonance (EPR) studies. The size and direction of the principal values of the electronic g-tensor (Table 3) indicated that the ligand-field environment of the heme iron was predominantly cubic, but contained components of lower symmetry. The assumption of rhombic symmetry (D_{2h}) for these weaker fields, where two principal g-values would be in the heme plane and coincide with two perpendicular axes pointing at the porphyrin nitrogens (Fig. 2), seems to lead to valid conclusions about the azido complexes, even though recent experiments by *Helcké et al.* (*40*) indicated that the actual ligand field might deviate somewhat from rhombicity.

K. Wüthrich

Table 3. *Principal g-values and coefficients for the three Kramers doublets in azidoferrihemoglobin (Kotani (59))*

$g_x = 1.72$	$g_y = 2.22$	$g_z = 2.80$	
j	A_j	B_j	C_j
1	0.973	−0.209	−0.097
2	0.219	0.970	0.108
3	0.071	−0.126	0.990

In D_{2h} symmetry the degeneracy of the three t_{2g}-orbitals is lifted. The three resulting orbital singlets, each of which still has a two-fold spin degeneracy, are mixed through spin-orbit coupling, and combine into a set of three Kramers doublets which may be written (59).

$$\Psi_j^+ = A_j (d_{yz})^+ + iB_j (d_{xz})^+ + C_j (d_{xy})^-$$
$$\Psi_j^- = -A_j (d_{yz})^- + iB_j (d_{xz})^- + C_j (d_{xy})^+ \qquad j = 1, 2, 3 \qquad (6)$$

The superscripts stand for the spin part of the wave functions. The coefficients A_j, B_j, and C_j are taken to be real, and their values for azidoferrihemoglobin are given in Table 3. Fig. 24 shows an energy diagram for the three states.

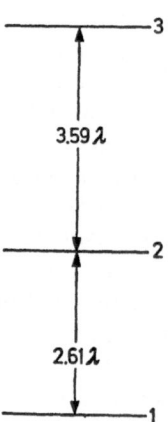

Fig. 24. Relative state energies (for hole functions) of the Kramers' doublets in azidoferrihemoglobin. The spin-orbit coupling constant λ is ca. 435 cm^{-1} in the free iron(III) ion, and might be smaller in the porphyrin complexes (95a)

Table 3 shows that each of the Kramers doublets corresponds closely to one of the orbitals d_{xz}, d_{yz}, and d_{xy}, to which small amounts of the other two t_{2g}-orbitals are admixed. From the energy separations between the three states (Fig. 24) one finds that the tetragonal splitting, which separates d_{xy} from d_{xz} and d_{yz}, is ca. 2060 cm^{-1}, if λ is taken as 435 cm^{-1}. The rhombic splitting, which separates d_{xz} from d_{yz}, would then be ca. 1040 cm^{-1}. At room temperature, where kT is ca. 200 cm^{-1}, the population probability would be small even for the first excited state in azidoferrihemoglobin.

Analysis of the structural features of the azide complexes led to the tentative conclusion that the azide ion was the major contributor to the anisotropy of the ligand field (*Weissbluth* (*106*); *Helcké et al.* (*40*)). One might then anticipate a decrease of the energy gaps between the different states, in particular between states 1 and 2 (Fig. 24), when going from the azides to other low spin compounds. Mössbauer experiments (*Lang* and *Marshall* (*67*)) and EPR studies (*Blumberg* and *Peisach* (*10*)) with polycrystalline samples of cyanoferrihemoglobin and cyanoferrimyoglobin gave indeed evidence that the energy separations are only ca. 1.0 to 1.4 λ between states 1 and 2, and 3.3 to 3.6 λ between states 1 and 3. This would correspond to appreciably smaller ligand field anisotropies than those observed in the azides.

To examine how the Kramers doublets are affected by a decrease of the ligand field anisotropy it is instructive to consider Ψ_1^+ (or $\Psi_{\bar{1}}$) as a wave function correct to first order, with the terms in d_{xz} and d_{xy} as the first order corrections (*106*). The spin-orbit coupling operator for a one-particle system is

$$)\text{-}(= -\lambda \vec{1} \cdot \vec{s} \tag{7}$$

We can then write

$$\Psi_1^+ = A_1 (d_{yz})^+ + \frac{\langle d_{xz}^+ | -\lambda \vec{1} \cdot \vec{s} | d_{yz}^+ \rangle}{\varepsilon_{yz} - \varepsilon_{xz}} (d_{xz})^+ + \frac{\langle d_{xy}^- | -\lambda \vec{1} \cdot \vec{s} | d_{yz}^+ \rangle}{\varepsilon_{yz} - \varepsilon_{xy}} (d_{xy})^-$$

$$= A_1 (d_{yz})^+ + \frac{i\lambda}{2 (\varepsilon_{yz} - \varepsilon_{xz})} (d_{xz})^+ + \frac{\lambda}{2 (\varepsilon_{yz} - \varepsilon_{xy})} (d_{xy})^- \tag{8}$$

ε_{xy}, ε_{xz}, and ε_{yz} are the hole orbital energies for either spin orientation, and hence $(\varepsilon_{yz} - \varepsilon_{xz})$ would be the rhombic splitting, and $1/2 \, [(\varepsilon_{yz} - \varepsilon_{xy}) + (\varepsilon_{xz} - \varepsilon_{xy})]$ the tetragonal splitting.

For the coefficients in Eq. (6) we further have that

$$A_j^2 + B_j^2 + C_j^2 = 1 \tag{9}$$

Eq. (8) then shows that the Kramers doublets will no longer correspond to almost pure t_{2g}-orbitals, but will contain appreciable admixtures of the other t_{2g}-orbitals if the tetragonal and rhombic splittings become considerably smaller than in the azides. Furthermore the population probability for the excited states would increase if the energy gaps between the three states became smaller (Fig. 24), and thermal mixing of the three Kramers' doublets might then also have to be considered.

D. Pseudocontact Shifts

In paramagnetic molecules with anisotropic g-tensors electron-proton dipole-dipole coupling may contribute to the hyperfine shifts observed in the proton NMR spectra. From the data to be discussed in this section it would seem, however, that in low spin ferric heme compounds many of the qualitative spectral features are mainly determined by Fermitype contact shifts.

If the conditions are known under which the dipolar terms are averaged in the solution, the pseudocontact shifts can be calculated from the molecular geometry and the principal values of the electronic g-tensor (*McConnell* and *Robertson* (76); *Jesson* (46)). This approach is used here to obtain an estimate for the pseudocontact shifts which might occur in low spin azidoferrihemoglobin. Assuming that only the ground state would be populated (1 in Fig. 24), and that the g-tensor would be axially symmetrical ($g_\parallel = g_z, g_\perp = g_x = g_y$), and neglecting possible effects of spin delocalization on the pseudocontact interactions the pseudocontact shifts for low spin ferric hemoproteins would be given by

$$\Delta \nu_{pct} = \frac{-\beta^2 \nu \, S \, (S+1)}{3 \, kT} \, (3 \cos^2 \chi_i - 1) \, \frac{1}{|r_i|^3} \, \frac{(g_\parallel + g_\perp) \, (g_\parallel - g_\perp)}{3} \qquad (10)$$

where β is the Bohr magneton, ν the proton resonance frequency, S the total electronic spin, $|r_i|$ the distance from the heme iron to the observed protons, and χ_i the angle between $\vec{r}_i$ and the $g_\parallel$-axis. If the tetragonal axis were perpendicular to the heme plane, and $g_\parallel > g_\perp$, $\Delta \nu_{pc}$ would shift the resonances of all the protons in the heme plane upfield. With $g_\parallel = 2.8$ and $g_\perp = 2.0$ these upfield shifts at 25 °C would be ca. 7.5 ppm for the mesoprotons, and ca. 2.5 ppm for the ring methyl protons, if $|r_i|$ were taken to be 4.5 Å and 6.4 Å, respectively (*Hoard* (41)). For the ring methyls these shifts would be small compared to the hyperfine shifts observed (Figs. 10—20). — In D_{2h} symmetry the pseudocontact shifts could be different for the individual mesoprotons and ring methyl groups. With a modified form of Equation (10) (*LaMar et al.* (65, 66)) one calculates that with $g_z = g_1 = 2.80$, $g_2 = 2.22$, and $g_3 = 1.72$ (Table 3) the resonances of all the protons in the heme plane would be shifted upfield by pseudocontact coupling. Depending on the orientation of the g-tensor in the molecule $\Delta \nu_{pc}$ for two of the mesoprotons might become comparable to the hyperfine shifts in the NMR spectra, but then the pseudocontact shifts of the other two mesoprotons would be rather small. For the four ring methyl groups $\Delta \nu_{pc}$ would still be comparatively small, and opposite in sign to the hyperfine shifts in Figs. 11 to 19.

Information on the relative importance of contact coupling and pseudocontact coupling was derived directly from examination of the NMR spectra. *Kowalsky (62)* concluded that the hyperfine shifts of the ferricytochrome c resonances at ca. -33 ppm and $+23$ ppm (Fig. 19) could not come from pseudocontact coupling. Otherwise, independent of the specific assignment of these lines, a considerable number of protons of the polypeptide chain would also experience sizeable pseudocontact shifts, and instead of the few resolved lines (Fig. 19) a rather large number of resonances would be observed in the spectral regions from -15 to -10 ppm, and 0 to $+5$ ppm. — In a different approach the resonances in ProtoCN and DeutCN were compared (Figs. 12 and 13). The similarity of the positions for the resonances of the mesoprotons (b) and the ring methyls (a) seems to indicate that the electronic structures are quite similar in the two molecules. Yet the resonances of the 2,4-protons in DeutCN (Fig. 12, f), and the α-protons of the 2,4-vinyls in ProtoCN (Fig. 13, f) are ca. 35 ppm apart, and the hyperfine-shifts of these two types of protons are opposite in sign. Since the positions relative to the heme iron, and hence $|r_i|$ in Eq. (10), are almost identical for these protons, this shift has to come essentially entirely from different contact coupling. A similar comparison of Mb^{III}CN and DeutMb^{III}CN (Fig. 17) leads to the same conclusion for the resonances in the hemoprotein molecule.

A more quantitative estimate of the pseudocontact shifts in DeutCN was obtained from comparison of the methylester resonances in DeutCN and deuteroporphyrin IX zinc(II) (114). Under identical conditions these resonances were both at -3.72 ppm from DSS in the diamagnetic complex, and at -3.78 and -3.82 ppm in DeutCN. The observed shift of ca. -0.08 ppm has to come almost entirely from pseudocontact coupling, since delocalization of unpaired spin density to these positions must be negligibly small, and a small change of the porphyrin ring current field, which might result from the zinc(II) vs. iron(III) substitution, would hardly be noticeable at the methylester positions either. With the assumption of an axially symmetrical g-tensor, and $|r_i|$-values for the different protons of 10.0, 6.4, 5.7, and 4.5 Å (*Hoard (41)*), one then estimates from the $|r_i|^{-3}$-dependence (Eq. 10) that the corresponding pseudocontact shifts would be ca. -0.3 ppm at the ring-methyl positions, -0.45 ppm at the 2,4-protons, and -0.8 ppm at the mesoprotons. These shifts would be very small compared to the hyperfine shifts observed in DeutCN.

The next example to be discussed is cyanoferriporphin. Fig. 10 shows that the resonances of the four mesoprotons are equivalent, and so are the resonances of the protons at positions 1 to 8. The importance of contact coupling is again quite obvious, since the hyperfine shift of ca. 24 ppm for the protons 1 to 8 is almost three times as large as that for the mesoprotons, which are nearer to the heme iron. Assuming that the unpaired electron density at the mesopositions is approximately 30% of that at positions 1 to 8 (*Longuet-Higgins et al.* (*68*)), and that $g_\parallel > g_\perp$, one finds that the pseudocontact shift for protons 1 through 8 could hardly be larger than a few percent of the observed total hyperfine shift. Similar assumptions, but with $g_\parallel < g_\perp$, lead to the same conclusion, because it seems that only an unreasonably large spin delocalization from the iron to the porphin ring could produce the observed spectrum if there were large downfield pseudocontact shifts.

As mentioned at the beginning of this section the size of the pseudocontact shifts in the NMR spectra could in principle be calculated for all the low spin ferric heme compounds if detailed data on the electronic g-tensors were available (*Jesson* (*47*)). Unfortunately the EPR data on the azides can not be used directly, because these complexes are not in a pure low spin state under the conditions of the NMR experiments (see section VI C). For the compounds in Figs. 10 through 20 no successful single-crystal EPR studies were as yet reported. However only g-values determined in frozen solutions are presently available (*Blumberg* and *Peisach* (*10*); *Salmeen* and *Palmer* (*95a*)), e.g. for dicyanoferriporphin at 1.4 °K $g_1 = 3.64$, $g_2 \approx 2.29$, and $g_3 \approx 1.0$ were found.

Since the magnetic equivalence of the individual ring protons (Fig. 10) would otherwise be lifted by pseudocontact coupling the principal values of the g-tensor have to be respectively perpendicular to the heme plane, and in the heme plane pointing at the pyrrole nitrogens. With g_1 along the z-axis (Fig. 3) the hyperfine shifts of the mesoprotons could then be accounted for entirely by pseudocontact coupling, which would also contribute ca. 20% of the shift observed for protons 1 to 8. Similar results are obtained for the compounds in Figs. 11 to 19. Yet the splitting of the individual mesoproton resonances into two groups of one and three lines (Figs. 11 to 13, 17) could not conceivably be caused by pseudocontact shifts. It is thus implied that there are still appreciable effects from contact coupling on the mesoprotons. As do some of the above-mentionned observations this then seems to imply that the pseudocontact shifts are smaller than what one calculates from the g-values in the frozen solutions, which could conceivably be influenced by lattice effects not present at ambient temperature.

A more detailed treatment, including the effects of spin delocalization on the pseudocontact shifts, might be warranted once single crystal EPR data will become available for several of the low spin ferric heme compounds. In the hemoproteins it would then be of special interest to investigate pseudocontact shifts for amino acid residues near the heme groups which could yield structural information in a similar way as the ring current shifts.

E. Contact Shifts

1. Electron-Proton Contact Coupling

In hemes and hemoproteins contact shifts arise if finite amounts of unpaired electron spin density are delocalized from the iron orbitals into the π-orbital systems of the porphyrin and the axial ligands, as indicated by the arrows in Fig. 25. Electron density is then further transferred from the aromatic ring carbon atoms to the protons (Fig. 2), thus giving rise to contact interactions. The measured isotropic contact coupling constants for the protons, A in Eq. (4), can be related to the integrated spin density on the neighboring ring carbon atom by (*McConnell (73)*; *Bersohn (8)*; *Weissman (107)*).

$$A = Q \, \varrho_C^\pi \qquad (11)$$

where Q is an empirical constant.

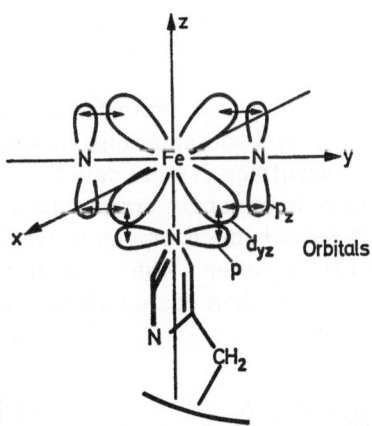

Fig. 25. Interactions of the $3d_{yz}$ ($3d_{xz}$) atomic orbital of the heme iron with the π-orbitals of the ligand nitrogen atoms. The arrows indicate likely pathways for electron spin delocalization from the iron to the ligands

For protons attached directly to the aromatic ring the spin transfer seems to be mainly by spin polarization (*McConnell* and *Chesnut* (*75*)) and $Q^H \approx -6.3 \cdot 10^7$ cps was found for a large number of aromatic compounds.

For protons of methyl groups attached to aromatic rings theoretical and experimental studies indicated that the contact interaction occurred mainly through hyperconjugation (*Bersohn* (*8*); *Bolton et al.* (*11*)). In a large number of aromatic radicals, including paramagnetic metal ion complexes, Q^{CH_3} was always positive, but its value was found to be quite different in different molecules (*Eaton et al.* (*27, 28*); *Horrocks et al.* (*44*)). An approximate value for Q^{CH_3} in low spin ferric hemes was derived from a comparison of the hyperfine shifts of the ring methyl protons and the 2,4-protons in DeutCN (Fig. 12). The similarity of the hyperfine shifts for the four ring methyls, as well as for the four meso-protons, indicates a rather symmetrical distribution of the spin density in the porphyrin ring of DeutCN. Therefore it appears that the average of the spin densities on the ring carbon atoms next to the four ring methyl groups should be approximately equal to the average of the spin densities next to the 2,4-protons. With this assumption, and using $Q^H = -6.3 \cdot 10^7$ cps, $Q^{CH_3} \approx 3.0 \cdot 10^7$ cps was obtained. Conclusions based upon comparative studies of different heme compounds should be little affected by possible small errors in Q, which could on the other hand strongly influence absolute measurements of the electron distribution between the iron and the ligands.

No generally applicable Q's can be given for methylene protons, for example of the propionates (Fig. 2), since the electron-proton coupling depends strongly on the sterical conformation of the molecule (*16*).

2. Spin Densities

Eq. (4), which relates the observed contact shifts of the proton resonances to their isotropic contact coupling constants, and hence to the spin densities on the ring carbon atoms, is valid only for systems with isotropic g-tensors. To obtain an estimate of the errors which might arise from its application to low spin ferric heme compounds, we shall briefly consider a more general form of the equation, which was given by *Jesson* (*46*) for tetragonal systems with more than one populated electronic state.

$$\Delta\nu_{ci} = -\frac{|\beta|\nu \, S' \, (S'+1)}{\hbar \, |\gamma_H| \, 3 \, kT} \sum_j \mathbf{f} \, (\tilde{g}, \tilde{A}_i)_j \, \frac{\exp\left(\dfrac{-E_j}{kT}\right)}{\sum\limits_j \exp\left(\dfrac{-E_j}{kT}\right)} \tag{12}$$

S' is an effective electronic spin (S' = 1/2), f is a function which depends on the averaging conditions in the solution, $\tilde{g}$ is the electronic g-tensor for the jth state, and $\tilde{A}_i$ the contact coupling tensor which describes the interactions of the ith proton with the electronic spin in the jth state. The summations are over the populated states, and E_j are the state energies (Fig. 24). Under the averaging conditions in solutions of hemoproteins, with $\tau_r \approx 10^{-8}$ sec, and $|g_{\parallel} - g_{\perp}| \geqslant 0.01$ at 220 Mc f is

$$f(\tilde{g}_i\tilde{A}_i) = \frac{g_{\parallel}A_{i\parallel} + 2g_{\perp}A_{i\perp}}{3} \tag{13}$$

The anisotropy of the contact coupling tensor was estimated from the relations (*Abragam* and *Pryce* (2); *Jesson* (46))

$$A_{i\parallel} = \tfrac{1}{2}g_{s\parallel} A_i \qquad g_{s\parallel} = 4\langle \Psi_j^+ | S_z | \Psi_j^+ \rangle$$
$$A_{i\perp} = \tfrac{1}{2}g_{s\perp}A_i \qquad g_{s\perp} = 4\langle \Psi_j^+ | S_x | \Psi_j^- \rangle \tag{14}$$

where $A_{i\parallel}$ and $A_{i\perp}$ are the parallel and perpendicular components of the axial tensor $\tilde{A}_i$, A_i is the isotropic contact coupling constant, and S_x and S_z are components of the electronic spin operator. For Ψ_j's similar to those of azidoferrihemoglobin (Eq 6 and Table 3) the anisotropy of $\tilde{A}_i$ was found to be very small. Therefore $f(\tilde{g},\tilde{A}_i)$ in Eq. (12) may be replaced by (g_0A_i), where $g_0 = \tfrac{1}{3}(g_{\parallel} + 2g_{\perp})$, and A_i is the isotropic coupling constant. Since g_0 in solution at ambient temperature is not known with certainty for any of the electronic states of the compounds studied (Figs. 10—20), an additional modification had to be made in the application of Eq. (12). From the principal g-values observed at low temperatures in the azides, and in polycrystalline samples of some of the cyanide complexes, it would appear that g_0 should not in general be greatly different from the free electron value. If we replace g_0 tentatively by $g_{\text{free electron}}$ Eq. (12) goes over into Eq. (4), with

$$A_i = \sum_j A_{ij} \frac{\exp\left(\frac{-E_j}{kT}\right)}{\sum_j \exp\left(\frac{-Ej}{kT}\right)} \tag{15}$$

The hyperfine coupling constant A_i would thus in general be temperature-dependent, as was indicated by some of the experiments.

In analogy to Eq. (15) the spin densities obtained with Eqs. (4) and (11) from the NMR shifts are the sum of the spin densities contributed by the different electronic states

$$\varrho_{Ci}^{\pi} = \sum_j \varrho_{Cij}^{\pi} \frac{\exp\left(\frac{-E_j}{kT}\right)}{\sum_j \exp\left(\frac{-E_j}{kT}\right)} \tag{16}$$

where i designates the ring carbon atom next to the ith proton, and the summations are over the populated states.

3. Molecular Orbitals

In the discussion of the Kramer's doublets we have dealt entirely with the properties of the atomic orbitals of the heme iron. In a different approach the electronic states are expressed in terms of molecular orbitals. This is particularly interesting for aromatic radicals because the unpaired electron spreads out over the entire conjugated system of π-orbitals.

Molecular orbitals φ_j may be written as linear combinations of suitable atomic orbitals Φ_i

$$\varphi_j = \sum_i c_{ij} \Phi_i \tag{17}$$

where the summation is over the atoms in the aromatic molecule, and c_{ij} are normalized coefficients. The probability that the unpaired electron density in the jth molecular orbital is on the ith atom is then $|c_{ij}|^2$. The observed spin densities ϱ_{Ci}^{π}, which correspond in this model to the sum of the spin densities in the different molecular orbitals (Eq. 16), can thus yield information on the coefficients of the atomic orbitals, and on the relative population of the different states in the molecular orbital description of the radical.

Hückel molecular orbitals in porphin were investigated by *Longuet-Higgins et al. (68)*, and the extended Hückel molecular orbital model was applied to metalloporphyrins in attempts by *Pullman et al. (93)*, *Ohno et al. (86)*, and *Zerner et al. (120)* to explain various experimental observations. Let us briefly consider a description of cyanoferriporphin. According to the Hückel theory all but the p_z-orbitals of each carbon and nitrogen atom of porphin are used up to form the relatively inert skeleton of single bonds. To describe the π-bonding twenty-four molecular orbitals of porphin can then be formed as linear combinations of the atomic p_z-orbitals. In D_{4h}-symmetry these orbitals belong to the irreducible representations A_{1u}, A_{2u}, B_{1u}, B_{2u} and E_g. Since it appears that of the iron atomic orbitals in low spin ferric hemes essentially only d_{xz} and d_{yz}, which belong to the Eg representation, contain unpaired electron density the analysis can be reduced to a study of the molecular orbitals formed by combination of the six Eg-orbitals of porphin with the d_{xz}- and d_{yz}-orbitals of the iron. The amounts of unpaired spin density delocalized in the different Eg-orbitals of porphin will depend on the Hückel parameters α_i and β_{ik} for all the different atoms, which may then be chosen to fit the experimental data.

In D_{4h}-symmetry the ring carbon atoms 1 to 8 (Fig. 2) are equivalent, and therefore the calculated spin densities are the same for all these atoms (*Longuet-Higgins et al.* (*68*)). The same holds for the four meso-positions. On the other hand, if a considerable rhombic splitting between d_{xz} and d_{yz} is introduced into the calculation, asymmetries will result between the spin densities along the x- and y-directions. For example a very large asymmetry would be present in a molecule with the orbital energies observed in azidoferrihemoglobin at low temperatures (*114*).

F. Molecular Structure and Electron Spin Delocalization

Table 4 lists the spin densities observed on the ring carbon atoms next to the ring methyls and the mesoprotons in some low spin ferric hemes and hemoproteins. It is seen that the average spin density on the porphyrin carbon atoms is somewhat smaller in the cyanoporphyrin iron(III) complexes than in the corresponding hemoproteins. It seems that ca. 20 to 35% of the unpaired electron are delocalized over the porphyrin ring in the different compounds. One observes further that the differences between the spin densities at the four ring methyl positions are much larger in the hemoproteins. More detailed insight into the relations between the molecular structures and these differences in the spin density distributions were obtained from studies of the dependence of the latter on the solvent, and on various modifications of the hemoprotein molecules.

1. Cyanoporphyrin-Iron(III) Complexes

The positions of the ring methyl and mesoproton resonances for cyanoferriporphin, MesoCN, and ProtoCN in different solvents are shown in Fig. 26. For all the three compounds the resonance positions show similar solvent dependences. This indicates that the solvent effects come probably from different solvation of the porphyrin ring, rather than from interactions with the side-chains. The changes with solvent are small compared to the overall hyperfine shifts. They are of the same order as the solvent dependences which were observed for the resonances in the diamagnetic porphyrin-zinc(II) complexes.

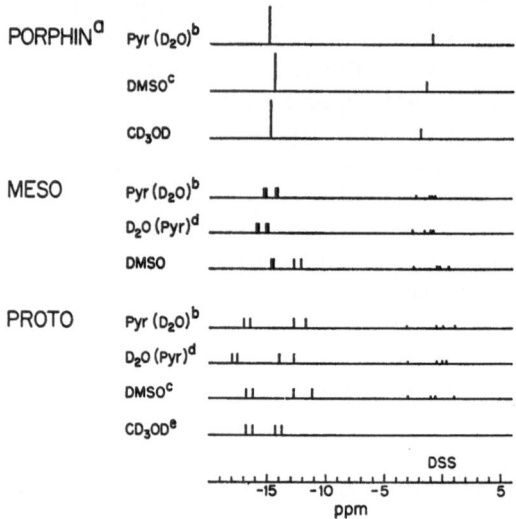

Fig. 26. Resonance positions for the ring methyls and mesoprotons of cyanoferriporphin, MesoCN, and ProtoCN in different solvents. a) The methyl resonance position was calculated from that observed for protons 1 to 8. b) d_5-pyridine/D_2O 3:1. c) Dimethylsulfoxide. d) d_5-pyridine/D_2O 1:3. e) the mesoproton resonances were not identified. T = 25 °C.

The spectrum of cyanoferriporphin corresponds to D_{4h} symmetry of the spin density distribution. This, and the small size of the changes with solvent, seem to confirm that two cyanide ions are bound at the axial positions of the heme iron (*Hogness et al. (42)*). Furthermore it seems from this spectrum that on the time scale of an NMR experiment complex formation with cyanide ion does not perturb the symmetry of the spin density distribution. On the other hand the equivalence of the positions 1, 3, 5, and 8 in cyanoferriporphin (Fig. 2) is lifted when the protons at positions 1 through 8 are substituted by various other groups in MesoCN and ProtoCN.

2. Hemoproteins

Comparison with Table 4 shows that the influence on the spin density distribution of the polypeptide environment in the hemoproteins is

Table 4. *Spin densities ϱ_C^{π} observed in some low spin ferric hemes and hemoproteins*

	cyanoferri-porphin a,b)	MesoCN[b])	ProtoCN[b])	Mb[III]CN[c])	Cytc[IIIc,d])
ring-CH$_3$	1.53	1.47	1.70	2.95	3.95
ring-CH$_3$	1.53	1.46	1.63	1.86	3.59
ring-CH$_3$	1.53	1.38	1.17	1.18	−0.74
ring-CH$_3$	1.53	1.37	0.95	1.12	−0.79
meso-H	0.58	0.58	0.68	1.21	
meso-H	0.58	0.57	0.62	0.78	
meso-H	0.58	0.55	0.59	0.75	
meso-H	0.58	0.48	0.43	0.69	

The spin densities on the ring carbon atoms next to the four ring methyl groups and the four mesoprotons were derived with the assumption that the observed hyperfine shifts came entirely from contact coupling. Eq. (4) and (11) were used, with $Q^H = -6.3 \cdot 10^7$ cps, and $Q^{CH_3} = 3.0 \cdot 10^7$ cps. ϱ_C^{π} is given in percent of one electron. For meso-H ϱ_C^{π} may be much smaller than the upper limits given here.

a) In place of the ring methyls the protons at positions 1 to 8 were observed (Fig. 10).
b) In a mixed solvent d_5-pyridine/D$_2$O 3:1.
c) In 0.1-M phosphate buffer, pD 7.0.
d) The ring methyl resonances were tentatively assigned, as described in the text. The mesoprotons have not been assigned.

considerably greater than the solvent effects in Fig. 26. It was of interest to locate the regions in the proteins which were involved in these heme-polypeptide interactions. For example the NMR spectra of myoglobins from different species, where a certain number of the amino acids are substituted due to evolutionary mutations, and of chemically modified myoglobins were studied. In these experiments the substitution or modification of amino acid residues outside the van der Waals distance from the porphyrin ring had no noticeable influence on the spin density distribution (*115*). Similar comparative studies of cytochromes c indicated again that only modification of the protein conformation in the immediate neighborhood of the porphyrin ring might noticeably affect the spin delocalization in the heme group (*111*).

Of the amino acids near the heme groups those bound to the axial positions of the heme iron appear to be most important, as one might have anticipated. The resonance positions for Mb[III]CN, DeutMb[III]CN, MesoMb[III]CN, and Cytc[III]CN, which all have a histidyl residue and a cyanide ion at the axial positions, were found to be quite similar, even though there are four different types of substituents at the 2- and 4-positions of the heme groups, and the polypeptide chains of the different proteins are also quite different. On the other hand the spectra of these four molecules are markedly different from those of ProtoCN, MesoCN, and DeutCN, which all have two cyanide ions at the axial positions, and of ferricytochrome c, where the axial ligands are histidine and methionine.

The observed dependence on the axial ligands seems to indicate that the asymmetry of the spin distribution in the hemoproteins could come mainly from interactions with the orbitals localized on the iron. A possible explanation might be that the axial histidyl residue interacts differently with the d_{xz}- and d_{yz}-orbitals of the iron. π-bonding interactions between the imidazole ring of the axial histidine and either d_{xz} or d_{yz} (Fig. 25) could affect the relative energies of these two iron orbitals. Thus the near-degeneracy of these two orbitals, which was quite apparent in the isolated hemes could be removed. Similarly in cytochrome c the sulfur atom of the methionine could form π-bonds with one of the d_{xz}- or d_{yz}-orbitals. Since the sulfur atom would contribute a p-orbital with a lone pair of electrons to this bond, a qualitative molecular orbital model would seem to imply that the most favorable configuration might arise if the histidine and the methionine were both bound to either d_{xz} or d_{yz}. From this one would expect a rather large asymmetry of the x- and y-directions, as it was observed in the NMR spectrum of ferricytochrome c (Fig. 19).

Additional information on the spin density distribution might be obtained from investigations of the proteins in H_2O rather than D_2O. Recent experiments reported by *Sheard* (*97*) showed that the resonances of some of the exchangeable protons, i.e. protons which are exchanged against deuterium in the deuterated solutions, are largely shifted by hyperfine interactions.

The conclusions derived from the NMR spectra were so far based mainly on the observation of the relative positions of the four ring methyl resonances, and the four mesoproton resonances of the heme groups. More detailed information on the structures in some of the molecules might be obtained if the resonances of the individual mesoprotons or ring methyl groups could be distinguished. A direct way to accomplish this in future experiments would be to study selectively deuterated heme groups, which might then also be introduced into reconstituted hemoprotein molecules. In the past evidence was obtained for a tentative assignment of the ring methyl resonances at -12.1 and -13.1 ppm in cyanoferrimyoglobin (Fig. 14). Comparison of Mb^{III}CN from sperm whale and porpoise (*115*), and investigation of the cyclopropane derivative of Mb^{III}CN (*Shulman et al.* (*100*)) indicated that these resonances come probably from the ring methyls at positions 1 and 5 of the heme group (Fig. 2).

V. Identification of the Axial Ligands of the Heme Iron in Cytochrome c

The iron of the heme group in cytochrome c is coordinated to the four nitrogens of the porphyrin (Fig. 19), and to two amino acid residues of the polypeptide chain. Early investigations indicated that the two axial ligands were bound to the iron through a nitrogen or sulfur atom, and the resulting "hemochromogen" spectrum of the heme group has long been known (*Keilin (50)*). On the other hand rigorous identification of these axial ligands has proved difficult, and some investigators even suggested that the ligands might not be the same in the two oxidation states. Of the potential hemochrome-forming amino acids, histidine, lysine, arginine, and methionine were most often cited as possible axial ligands. Reviewing the research on this problem *Margoliash (71 a)* concluded that the histidyl residue at position 18 of the polypeptide chain, and methionyl at position 80 are the most probable axial ligands in reduced cytochrome c. It was left open whether the same ligands are bound in both oxidation states. More recently X-ray studies by *Dickerson et al.* (25, 26) at 4 Å resolution showed that the axial ligands in horse ferricytochrome c are histidine-18 and an aliphatic amino acid residue, which could be methionine-80. Improvement of the resolution should soon identify with certainty the sixth ligand in crystalline ferricytochrome c. NMR studies were found to be particularly suitable for studies of the solution structures. They provided good evidence that the sixth ligand is methionine in reduced and oxidized cytochrome c, and that both axial ligands are identical in the two oxidation states. This leaves the possibility open, that the redox reaction in cytochrome c might occur without major changes of the protein conformation, whereas it would appear that interconversion between the ferric and ferrous oxidation states would necessarily induce appreciable conformational changes if one of the axial ligands were different in the two states (71).

The NMR investigations involved studies of the spectra of ferrocytochromes c, ferricytochromes c, and the complexes with cyanide ion. For the interpretation of the data it was particularly important that methionine is the only potential hemochrome-forming amino acid which contains a methyl group. Since some of the arguments used might also apply to NMR studies of metal-ion coordination in other biological molecules a rather detailed account of these experiments is presented.

A. Ferrocytochrome c

Reduced cytochrome c is diamagnetic (Table 1). The NMR spectrum, which was first reported by *McDonald* and *Phillips* (77), consists of the strongly overlapping resonances of ca. 650 protons in the spectral region from -0.5 to -10 ppm, and some resolved lines at high fields from DSS. In the high field region there are resonances of intensity three protons at 3.3, 0.7, 0.6 and 0.6 ppm, a resonance of two protons at -0.1 ppm, and one-proton resonances at 3.7, 2.7, 1.9, 0.2, and -0.2 ppm. All but one of the line positions are independent of temperature (Fig. 27). One methyl resonance moves from 0.6 ppm at 35 ° to 0.7 ppm at 12 °, indicating localized changes of the protein conformation in this temperature range.

The methyl resonances of aliphatic amino acids are usually observed at around -1.0 ppm, those of the methylene groups at somewhat lower fields (78). Hence the ferrocytochrome c lines between 2.7 and 3.7 ppm have to be shifted approximately 4 to 5 ppm upfield by interactions with neighboring groups in the molecule. One would expect an upfield shift of this order for some of the protons of the axial ligands of the heme iron

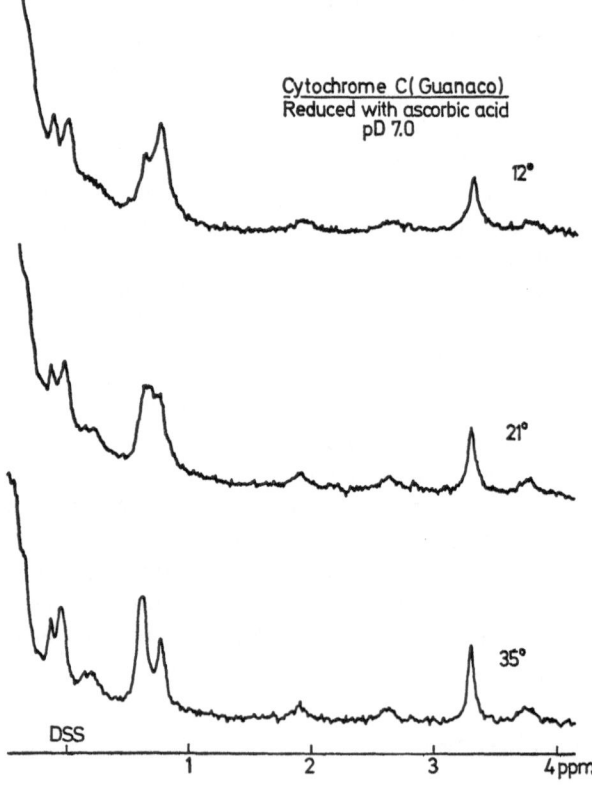

Fig. 27. Dependence on temperature of the spectral region from -1 to 4 ppm in the NMR spectrum of ferrocytochrome c. (Reproduced from ref. (111))

(Fig. 7). Similar shifts might result if several rings of aromatic amino acids were to form a "cage" around a group of protons which would then be subjected to all these ring current fields simultaneously. However, since these large ring current shifts were observed only in ferrocytochrome c and not in ferricytochrome c (110), the assignment to one or both axial ligands appeared most probable. Concluding that the methyl resonance at 3.3 ppm (Fig. 27) indicates that one of the axial ligands is methionine, *McDonald et al.* (79) further suggested that the three one-proton lines at 3.7, 2.7, and 1.9 ppm correspond to three protons of the γ- and β-methylenes of the axial methionyl residue.

The resonances between —0.2 and 0.7 ppm could be shifted to their high field positions if the corresponding protons were located near the heme outside the area of the axial ligands (Fig. 7), or near any one of the aromatic amino acids. *McDonald et al.* (79) tentatively assigned the three methyl resonances to the threonyl residue at position 19 of the polypeptide chain, and to isoleucine-81.

B. Ferricytochrome c

The discussion in section IV A indicated that the high field resonance at +23.2 ppm of ferricytochrome c (Fig. 19) comes most probably from one of the axial ligands. The intensity of this resonance corresponds to three protons, and hence one of the axial ligands in oxidized cytochrome c would then seem to be methionine. Additional support for this assignment was obtained from studies of the width of the proton resonances.

The line-broadening from proton-proton dipolar coupling should be quite similar for the ring methyl resonances of the heme group, and the methyl resonance of an axial methionyl residue. The line-broadening from electron-proton scalar coupling can be estimated through Eq. (5). With $T_{1e} = 2 \cdot 10^{-12}$ sec one finds that a ring methyl resonance shifted to —33 ppm at 35 °C (Fig. 19) should be broadened by ca. 8.5 cps, and the methyl resonance of an axial methionine at +23.2 ppm by ca. 3.8 cps. A difference of this order would hardly be noticeable in the spectra. On the other hand appreciably different line-widths result from dipole-dipole coupling between the electron and the protons. The average distance from the heme iron is ca 6.4 Å for the ring methyl protons of the heme group (41), and ca. 3.6 Å for the methyl protons of an axial methionine. Inserting these values for r_i, and $\tau_c = T_{1e} = 2 \cdot 10^{-12}$ sec into the second term of Eq. (5), one finds that at 220 Mc the line-broadening from electron-proton dipolar coupling should be ca. 1.1 cps for the ring methyls, and ca. 36 cps for the methyl resonance of methionine. The spectrum of Fig. 19 indicates indeed that the resonance at 23.2 ppm is appreciably broader than the other resolved methyl resonances. Eq. (5) would further predict that the line-broadening from electron-proton dipolar coupling should increase at lower fields, whereas in the region considered the other two major relaxation mechanisms are to a good approximation independent of the field strength. At 100 Mc the dipolar line-broadening should be ca. 1.7 cps for the ring methyls, and ca. 55 cps for the methyl resonance of methionine. In qualitative agreement with these calculations measurements on a Varian XL-100 spectrometer[4]) gave a full width at half height of the resonance at +23.2 ppm of ca. 100 cps, as compared to ca. 30 cps for the resonances at −31 and −34 ppm.

C. Cyanide Complexes of Cytochrome c

The hyperfine shifts in the NMR spectra of ferricytochrome c and its cyanide complex (*Horecker and Kornberg* (*43*)) are markedly different (*110*), so that the complex formation can be observed in the NMR spectrum (Fig. 28). It is found that a 1:1 complex is formed.

At high fields the NMR spectrum of cyanoferricytochrome c contains resolved hyperfine-shifted lines at +1.2 and +4.4 ppm (Fig. 29). Upon addition of dithionite these lines, as well as the resonances between −10 and −30 ppm, disappear immediately, indicating fast reduction to the diamagnetic ferrous oxidation state. The primary reaction product has no proton resonances at higher field than +1 ppm. However, with time elapsing after the reduction the NMR spectrum of ferrocytochrome c appears slowly. On the other hand the entire ferrocytochrome c spectrum was observed immediately after reduction of ferricytochrome c.

These observations were explained by a mechanism proposed by *George* and *Schejter* (*34*): Under the conditions of the experiment in Fig. 29 the reduction of cyanoferricytochrome c to the ferrous oxidation state is fast, and is followed by slow dissociation of the unstable cyanoferrocytochrome c. In the course of this dissociation cyanide ion is replaced by the axial ligand of native ferrocytochrome c. This experiment implies even more strongly than the mere observation of the ferrocytochrome c spectrum that this axial ligand is probably methionine. It would seem very unlikely that upon dissociation of cyanoferrocytochrome c the protons corresponding to the four ferrocytochrome c resonances between 1.9 and 3.7 ppm could move simultaneously into locations with extremely high ring current fields, if it were not at one of the axial positions of the heme iron.

---

[4]) We thank Dr. *U. Scheidegger* from Varian A. G. in Zürich who made the measurements on this instrument possible.

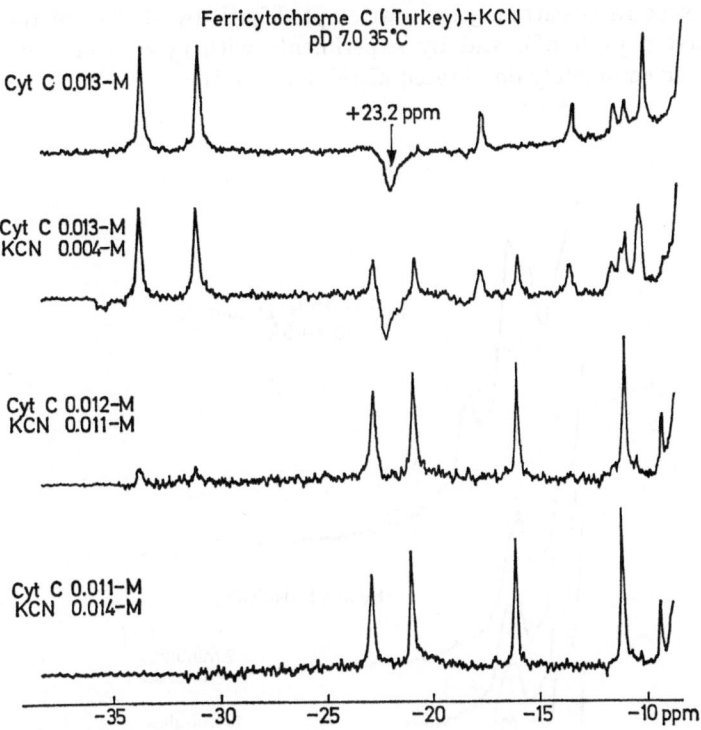

Fig. 28. NMR study of the complex formation of ferricytochrome c with cyanide ion. No further spectral changes were observed upon addition of an excess of KCN. (Reproduced from ref. (*111*))

From temperature studies similar to that described for cyanoferri-myoglobin (Figs. 15 and 16) the ring current-shifted high field resonances of ferricytochrome c and cyanoferricytochrome were identified (*110, 111*). The methionyl resonances between 1.9 and 3.7 ppm were not observed for either of the ferric compounds. This was to be expected, since in ferricytochrome c these lines are further shifted by hyperfine interactions, and in cyanoferricytochrome c the methionine is no longer one of the axial ligands.

Since methionine seems to be bound in both oxidation states, and yet is the more labile axial ligand in the reaction with cyanide ion, the NMR data seem to support that histidine is also an axial ligand in ferri- and ferrocytochrome c. Direct identification of the axial histidine in the

NMR spectra is currently being attempted both by studies of the fully protonated protein[5]), and by experiments with cytochromes c which contain a completely deuterated histidyl-18 residue.

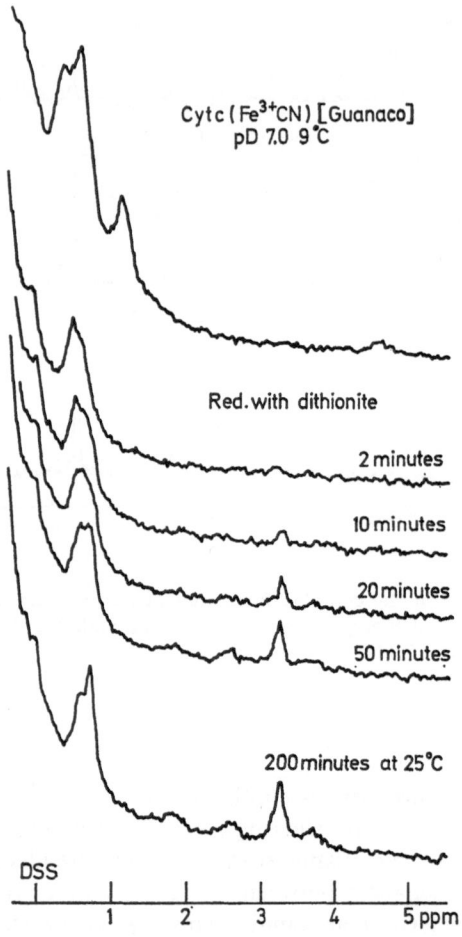

Fig. 29. NMR spectrum between -1 and $+5$ ppm of cyanoferricytochrome c, and of the reaction products observed at variable times after reduction of cyanoferricytochrome c with dithionite (Reproduced from ref. (*111*))

[5]) *B. Sheard* and *R. G. Shulman*, private communication. *A. Redfield*, private communication.

D. Cytochromes c from Different Species

Cytochromes c from different species contain the same heme group but may differ in the nature of the amino acid residues at various positions of the polypeptide chain. The functional importance of those segments of the protein which are invariant among different species has long been recognized (*Margoliash* and *Schejter* (*71*)). The NMR spectra in the reduced and oxidized forms indicate that methionine is an axial ligand in all the mammalian-type cytochromes c studied so far (*McDonald et al.* (*79*); *Wüthrich* (*111*)).

Methionine-80 is the only invariant methionyl residue among the species studied, and hence would seem to be the most likely axial ligand. Evidence for this assignment was also obtained from the X-ray studies by *Dickerson et al.* (*25, 26*). As position 80 is part of the longest invariant segment of the polypeptide chain the present data on the coordination of the heme iron seem to confirm the importance of this invariant region.

VI. Myoglobin and Hemoglobin

In addition to the low spin ferric states of myoglobin and hemoglobin (Section IV) the high spin ferric, and the high and low spin ferrous compounds have also been studied. It was of particular interest to investigate how the biologically important interconversions between the unligated and ligated ferrous forms of myoglobin and hemoglobin affected the molecular structures in solution.

A. Ferrous Myoglobin

1. Proton Magnetic Resonances of Oxymyoglobin and Deoxymyoglobin

The proton NMR spectrum of oxymyoglobin is shown in the lower part of Fig. 30. Even though the individual resonances overlap rather strongly some spectral features can be recognized, and there are two well-resolved lines of relative intensities one and six at +2.8 and +0.1 ppm. Very similar spectra were observed after replacement of O_2 with CO or with ethylisocyanide (*Shulman et al.* (*101*)).

In the NMR spectrum of deoxymyoglobin some resonances at low fields from —10 ppm (Fig. 31) and at high fields from DSS were found to be shifted by hyperfine interactions with the paramagnetic heme iron. There are fewer resolved lines than in cyanoferrimyoglobin (Fig. 14),

103

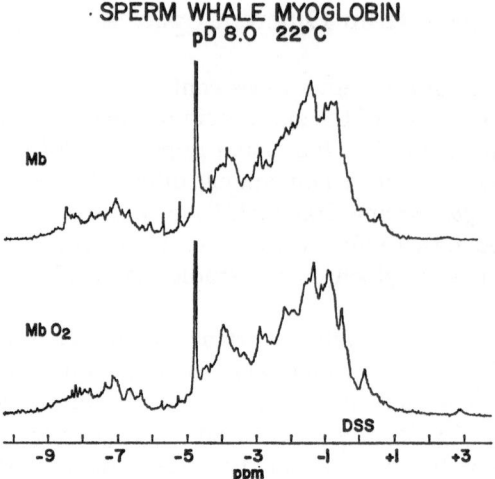

Fig. 30. Proton NMR spectra between +4 and −10 ppm of deoxymyoglobin (Mb) and oxymyoglobin (MbO₂) from sperm whale

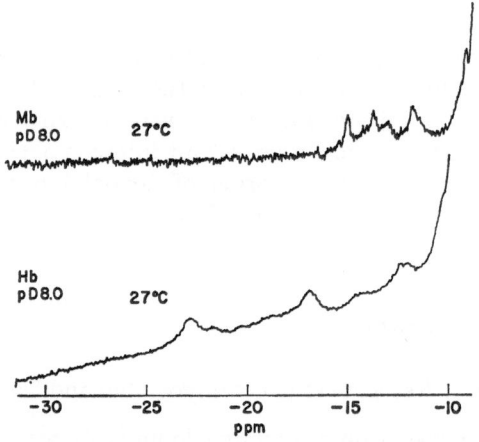

Fig. 31. Proton NMR spectra between −10 and −35 ppm of deoxymyoglobin and deoxyhemoglobin. The Hb^{II}-spectrum was recorded after time-averaging for 10 hours with a Fabritek-1070 computer

indicating that all the heme resonances had not been observed. From their intensities the lines in Fig. 31 were tentatively assigned to the ring methyls of the heme group. Since these resonances are not very broad it seems most probable that the remaining lines have not been observed because they were not shifted outside of the diamagnetic region. It is interesting that the hyperfine shifts in deoxymyoglobin appear to be considerably smaller than in cyanoferrimyoglobin, even though the total electronic spin is larger. This might be related to differences between the molecular structures of the heme groups in the two electronic states. For example if the iron were located further out of the porphyrin plane in the high spin ferrous state one might expect relatively small hyperfine shifts.

In the spectrum of deoxymyoglobin most of the spectral features between DSS and −10 ppm, and three resonances of intensities three protons at +0.6, +0.3 and 0 ppm (Fig. 30) were found to be independent of temperature, and hence seem to be determined essentially by the protein conformation in the molecule (*Shulman et al. (100, 101)*).

2. Computer Simulation of the Spectra

The interactions between the different amino acid residues in the three-dimensional structure of a protein molecule (Fig. 1) displace the individual resonances as indicated in Fig. 8. The resulting shifts are sufficient to produce spectral features which are markedly different from those of a hypothetical polypeptide spectrum obtained by addition of the individual amino acid resonances. In principle the calculation of the protein spectrum from the spectra of the individual amino acids would be possible if the amino acid composition, the three-dimensional structure in solution, and the local magnetic fields of all the groups in the molecule were known. However valuable information may be obtained through a less ambitious effort based on the known amino acid sequence, the atomic coordinates obtained by X-ray studies in single crystals, and the local magnetic fields of the aromatic rings in the molecule (*101*). Since in a diamagnetic protein the aromatic ring current fields are by far the largest local magnetic fields this calculation should in favorable cases provide evidence for the assignment of the resolved resonances at high fields, provided that the molecular solution structure is similar to that in single crystals of the protein (*Hügli* and *Gurd (45)*). A more detailed simulation which would include interactions with the local fields of all the different functional groups in the molecule seems at present hardly possible, since the strengths of most of these local magnetic fields are rather difficult to estimate (*Sternlicht* und *Wilson (103)*).

In paramagnetic hemoproteins dipole-dipole coupling between the protons of the polypeptide chain and the unpaired electrons of the heme iron may also affect the appearance of the spectral region from −0.5 to −10 ppm. These interactions will in general produce pseudocontact shifts (Eq. 10), and broadening of the proton resonances (Eq. 5). This can also be included in the partial computation of the NMR spectra. In low spin ferric compounds, where T_{1e} is extremely short, the line widths of the polypeptide resonances will be essentially unaffected by the presence of the paramagnetic heme iron, whereas considerable line-broadening would be expected in ferrimyoglobin.

3. Structural Changes upon Oxygen Binding

A comparison of the two spectra in Fig. 30 shows that there are a sizeable number of differences between Mb^{II} and $Mb^{II}O_2$. Even though it could not be excluded that some of the differences might be related to the paramagnetism of Mb^{II} careful examination of the temperature dependence of the resonances implied that most of the differences come from changes of the protein conformation (*Shulman et al. (101)*). In particular the different positions of some of the resolved ring current shifted lines in the spectral regions from -0.3 to $+3.0$ ppm and from -6.5 to -5.0 ppm (Fig. 30), where there are no resonances of the individual amino acids, seem to be related to structural changes near the heme group. Comparison of the observed resonances with the computed resonance positions indicated that the spectral changes upon oxygen binding would correspond to variations of the distances between the heme iron and some nearby amino acid residues (e.g. phe CD-1, val E-11, leu B-10, see Fig. 1) of at least 0.2 Å.

B. Ferrimyoglobin

In ferrimyoglobin the sixth position of the heme iron is occupied by a water molecule (Fig. 3), and the molecule is in a high spin configuration (Table 1). As for cyanoferrimyoglobin (Fig. 14) the proton NMR spectrum consists of the strongly overlapping polypeptide resonances between DSS and -10 ppm, and the hyperfine-shifted lines of the heme protons. However, because of the larger total electronic spin the hyperfine shifts are considerably larger, and dipolar broadening of the proton resonances is much greater than in the low spin ferric compounds.

Kurland et al. (64) observed hyperfine-shifted resonances of ferri-myoglobin between -20 and -90 ppm. Comparison with the spectra of the iron(III)-complexes with protoporphyrin IX, deuteroporphyrin IX, and mesoporphyrin IX seems to indicate that four of the low field resonances correspond to the four ring methyl groups. Additional lines at high fields from DSS were observed in ferrimyoglobin (*116*), and in high spin porphyrin-iron(III) complexes (*Caughey et al. (19)*). The observation of hyperfine shifts to high and low fields, and the large size of these shifts seem to indicate that the electron spin delocalization in high spin ferric heme groups might occur via π- and σ-type wave functions. It should be possible in the near future to extend the NMR investigations to a larger number of high spin ferric hemes and hemoproteins, and to study in detail the influence of the axial ligands on the electronic structure (*Caughey et al.* (*19, 19a*)).

The observation of the hyperfine-shifted resonances in high spin ferric hemes and hemoproteins is more difficult than in the low spin ferric compounds because of line-broadening effects. This implies that the longitudinal electronic relaxation time is considerably longer than in the low spin ferric state. Therefore the dipolar interactions with the heme iron should also broaden a considerable number of the resonances of the polypeptide chain in ferrimyobglobin, whereas there should be no pseudocontact shifts at ambient temperature (*Kurland et al.* (*64*)). From comparison of the spectral region between DSS and -10 ppm with the spectra computed with inclusion of dipolar broadening for different values of T_{1e}, the electronic relaxation time was estimated to be ca. $1 \cdot 10^{-10}$ sec (*116*). This value appears to agree quite well with observations from relaxation enhancement measurements[6]. A line width of several hundred cps for the ring methyl resonances, which would correspond to $T_{1e} \approx 1 \cdot 10^{-10}$ sec (Eq. 5), would seem to be compatible with the experimental observations (*64*).

C. Azidoferrimyoglobin

From measurements of the magnetic susceptibilities in solutions of azidoferrimyoglobin *Beetlestone* and *George* (*7*) concluded that the electronic configuration in this compound is a thermal mixture of high spin $(S = 5/2)$ and low spin $(S = 1/2)$ states. The NMR data seem to be compatible with this interpretation.

The basic features of the $Mb^{III}(N_3)$ spectrum are similar to those of cyanoferrimyoglobin (Fig. 14). Only one set of hyperfine-shifted resonances is observed, but the hyperfine shifts are larger and the lines broader than in $Mb^{III}CN$. Furthermore the temperature dependences of the resonances positions show drastic deviations from Curie's law (Eq. 4). Similar data were also obtained for azidoporphyrin-iron(III) complexes in pyridine solution (*116*).

From the observed resonance positions for $Mb^{III}(H_2O)$ $(S = 5/2)$ and $Mb^{III}CN$ $(S = 1/2)$ one might estimate that at 220 Mc two sets of

---

[6] *S. H. Koenig*, private communication.

hyperfine-shifted resonances, which would correspond to the high and low spin species in solutions of $Mb^{III}(N_3)$, should have been observed if the rate of interconversion between the two forms were slower than ca. $1 \cdot 10^5 \sec^{-1}$. On the other hand in the limit of fast exchange the hyperfine shifts for the $Mb^{III}(N_3)$ resonances should be the weighed sums of the shifts for the low spin and high spin forms, which may be given approximately by

$$\Delta\nu[Mb^{III}(N_3)] = \alpha \, \Delta\nu[Mb^{III}CN] + (1 - \alpha)\Delta\nu[Mb^{III}(H_2O)] \qquad (18)$$

where α is the fraction of low spin form present, and $\Delta\nu$ are the hyperfine shifts for corresponding resonances of the three compounds. For $\alpha = 0.78$ (7), and the ring methyl resonance positions observed in $Mb^{III}CN$ (Fig. 14) and $Mb^{III}(N_3)$ (116), the hyperfine shifts for $Mb^{III}(H_2O)$ calculated with Eq. (18) agreed closely with observed resonance positions (Fig. 2 of ref. (64)).

The resonance positions in solutions of $Mb^{III}(N_3)$ were found to be independent of the excess of azide ion added. This would appear to indicate that the equilibrium between high and low spin states is not directly related to the extent of the complex formation with azide ion. It appears rather that there might be a rapid equilibrium between $Mb^{III}(N_3)$ (S = 1/2) and $Mb^{III}(N_3)$ (S = 5/2).

D. Hemoglobin

Investigations by various techniques provided evidence that the molecular structures of deoxyhemoglobin and oxyhemoglobin show significant differences (*Perutz et al.* (80, 90); *Guidotti* (39); *Ogawa* and *McConnell* (84, 85)). The question was left as to how the conformational changes during oxygenation might be related to the cooperativity of ligand binding evidenced by the shape of the oxygenation curve (Fig. 4). The observation of the proton magnetic resonances of cyanoferrihemoglobin (Fig. 18), and various other derivatives of hemoglobin, provided a means by which information on the cooperativity of ligand binding could be obtained.

The NMR spectra for the different electronic configurations described for myoglobin were also studied for hemoglobin and for the isolated α- and β-chains (*Shulman et al.* (99)). The basic spectral features are similar to those of the corresponding myoglobins, but the size of the hyperfine shifts of the heme resonances is quite different. As an illustration the resonances at low fields of deoxymyoglobin and deoxyhemoglobin are compared in Fig. 31.

As was pointed out by *Perutz* (*89*) the protein structures of myoglobin and of the individual hemoglobin subunits seem to be quite similar, but recent investigations indicated that the subunit structures are less rigid both in single crystals (*McLachlan et al.* (*80*)) and in solution (*118*). The implication is that conformational changes corresponding to those observed in myoglobin, or even greater, should occur upon ligand binding to the individual subunits of hemoglobin. The question then arises as to how far these changes of the tertiary structure would extend from the ligated subunit into the neighboring subunits (Fig. 32). Since the subunit-interactions affect the binding of the ligands, which occurs at the heme groups, it was of particular interest to investigate if conformational changes could be detected at the sites of the neighboring hemes.

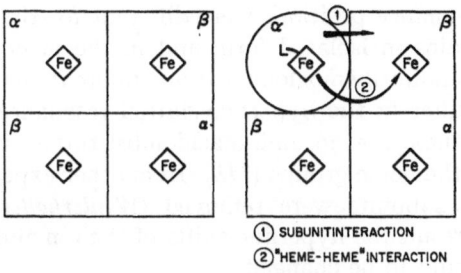

① SUBUNITINTERACTION
②"HEME-HEME"INTERACTION

Fig. 32. Schematic representation of a hemoglobin molecule. Ligand binding changes the conformation of the ligated subunit, and might affect the ligand affinity of the neighboring subunits by different mechanisms

Comparison of the diamagnetic regions of the NMR spectra of Hb^{II} and $Hb^{II}O_2$ showed quite obvious differences (*Shulman et al.* (*99*)), indicating differences between the protein conformations. However, because of the strong overlap of the resonances in this region (Fig. 18) less detailed information was obtained than for myoglobin. It was not possible to distinguish between spectral features which might be related to the quarternary and tertiary structures. Furthermore it would not generally be possible from the polypeptide resonances in a partially ligated hemoglobin (Fig. 32) to distinguish between spectral features related to the ligated subunits, and to the neighboring chains. On the other hand studies of the hyperfine-shifted resonances of the heme groups made it possible in certain cases to distinguish between the individual subunits.

1. Hyperfine Shifts in Hemoglobin

In deoxyhemoglobin and cyanoferrihemoglobin the resonances of the α- and β-subunits were found to be somewhat different (*Shulman et al.* (*99*)). This indicated again that the hyperfine shifts of the heme resonances are sensitive to the heme environment (see Figs. 10 to 20, and 31). Hence if ligation of one or several of the hemoglobin subunits produced changes of the hyperfine-shifted resonances in the neighboring unligated hemes one might conclude that the ligand-induced conformational changes were propagated to the positions of the heme groups in the neighboring subunits.

Experimental evidence was found that such "long-range" conformational changes could be detected by the NMR technique. For example, even though the hemes are not located on the interfaces, the interactions between the hemoglobin subunits seem to influence their protein environment. This was evidenced by the observation of *Shulman et al.* (*99*) that the resonance positions were different for the heme protons of the α- and β-chains in isolated form and in the hemoglobin tetramer. Furthermore cyanoferrihemoglobins from different mammalian species were found to differ in the hyperfine-shifted resonances of their NMR spectra even if there are no amino acid substitutions in the immediate surroundings of the heme groups (*118*). In another experiment the heme groups of the β-subunits were removed (*Winterhalter* and *Deranleau* (*108*)), and as a result the hyperfine-shifts of the remaining hemes of the α-chains were found to be changed.

2. Subunitinteractions in Hemoglobin A

The oxygenation curve of hemoglobin (Fig. 4) indicates that the unligated subunits in a partially ligated hemoglobin molecule (Fig. 32) have a higher ligand affinity than completely unligated hemoglobin. Since the oxygenated chains are diamagnetic comparison of the hyperfine shifts in deoxyhemoglobin and in the partially ligated forms would show if the spin distribution in the heme groups was affected by the ligand-binding to the neighboring chains. Partially oxygenated hemoglobins have not yet been isolated in sufficient quantities for such a direct NMR investigation of possible relations between oxygen affinity and the spin distribution in the heme groups. On the other hand the effects of ligand binding on the neighboring heme groups were studied in a series of mixed-state hemoglobins e.g. $\alpha_2^{II}\beta_2^{III}(CN)$, $\alpha_2^{II}(O_2)\beta_2^{III}(CN)$, $\alpha_2^{II}\beta_2^{III}(H_2O)$, etc. So far in these model compounds no effects of ligand binding on the proton resonances of the neighboring heme groups were observed. In analogy to these data it was concluded that the increased oxygen affinity of partially oxygenated hemoglobin is probably not related to interactions which would extend from the ligated heme to the hemes of the neighboring subunits (*Shulman et al.* (*99*)). Furthermore the NMR data appear to imply, in a similar way as *Ogawa* and *McConnell's* (*84, 85*) spin labelling experiments, that the oxygenation of hemoglobin A could best be described by a sequential model (*Koshland et al.* (*58*)) for subunit interactions. Different detailed mechanisms might still be written. For example the conformational changes in the ligated subunits might affect the stability of the subunitinterfaces in the tetrameric hemoglobin molecule, and thus affect the oxygen affinity of the remaining unligated chains.

VII. Relaxation Enhancement Measurements

Structural information on hemoproteins can also be obtained from investigations of the enhancement of the nuclear spin relaxation in the bulk water. In this section a qualitative discussion is presented of relaxation mechanisms in solutions of diamagnetic and paramagnetic metalloproteins, followed by a brief survey of experiments with hemoproteins.

A. Relaxation Mechanisms

In protein solutions the water protons may be considered to reside in two different environments, i.e. the bulk water, and the hydration spheres of the protein molecules. If there is fast exchange of protons between these environments a single proton nuclear magnetic resonance will be observed, which corresponds to the average of the resonances in the different environments. Following *McConnell* (*74*) the observed longitudinal relaxation time is to a good approximation

$$T_1^{-1} = \frac{n[Pr]/[H_2O]}{T_{1pr} + \tau_{pr}} + T_{1w}^{-1} \tag{19}$$

$[H_2O]$ is the molarity of water in the protein solution, T_{1pr} and T_{1w} are the relaxation times for protons in the hydration spheres of the protein and in the bulk water, $[Pr]$ is the protein concentration in mol/l, n the number of waters bound to each protein molecule, and τ_{pr} the residence time for water protons on the protein molecule. In writing Eq. (19) it was assumed for simplicity that there is only one type of hydration site with characteristic τ_{pr} and T_{1pr}. It is seen that the relaxation enhancement through the presence of the protein is

111

determined by the T_{1pr}-relaxation or the chemical exchange of protons, whichever is slower. Furthermore the relaxation enhancement would be proportional to the protein concentration and the hydration number of the protein molecules.

Explicit expressions for T_{1pr} were derived from models for the interactions of a proton with the local magnetic fields of its environment. Two types of environments which are important in hemoproteins will be discussed. First, if the water molecule is in a diamagnetic environment, dipole-dipole coupling between the two protons modulated by the anisotropic rotational tumbling of the molecule will be the dominant relaxation mechanism. The resulting relaxation rate is (*Solomon (102); Abragam (1)*)

$$(T_{1pr}^d)^{-1} = \frac{3}{10} \frac{\hbar^2 \gamma_I^4}{r^6} \left\{ \frac{\tau_c}{1+(\omega_I\tau_c)^2} + \frac{4\tau_c}{1+(2\,\omega_I\tau_c)^2} \right\} \qquad (20)$$

where γ_I is the gyromagnetic ratio for protons, r the distance between the protons, and ω_I the resonance frequency for protons in radians/sec. For the effective correlation time we have that $1/\tau_c = 1/\tau_{pr} + 1/\tau_r$, where τ_r is the correlation time for rotational tumbling of the hydrated protein molecule. Expression (20) shows that the relaxation rate increases linearly with τ_c as long as $\omega_I\tau_c \ll 1$.

The second case to be discussed here is that of protons bound to the protein near the paramagnetic centers. The relaxation arising from dipole-dipole coupling with the unpaired electrons is given by (*Solomon (102); Abragam (1)*)

$$(T_{1pr}^p)^{-1} = \gamma_I^2 \gamma_S^2 \hbar^2 \frac{2\,S\,(S+1)}{15\,r^6} \left\{ 3\,\tau_c + \frac{7\,\tau_c}{1+(\omega_s\tau_c)^2} \right\} \qquad (21)$$

where γ_I and γ_S are the gyromagnetic ratios for the proton and the electron, S is the total electronic spin, and r the proton-electron distance. For the effective correlation time we have that $1/\tau_c = 1/\tau_{pr} + 1/\tau_r + 1/T_{1e}$, where T_{1e} is the longitudinal electronic relaxation time. In writing Eq. (21) the assumption was made that $\omega_I\tau_c \ll 1$, which should be valid for paramagnetic hemoproteins, since $T_{1e} \ll \tau_r$. Because $\gamma_S = 658\,\gamma_I$ the paramagnetic contribution to the overall relaxation can become much larger than that of Eq. (20). The dipolar relaxation decreases fast with increasing distance from the heme iron, and will further depend very sensitively on τ_c.

A detailed discussion of the contributions from different relaxation mechanisms to the observed relaxation rates in solutions of paramagnetic metal ions was presented by *Swift* and *Connick (104)*. *Pfeifer (92)* considered the situation in solutions of paramagnetic hemoproteins, and *Koenig* and *Schillinger* (*55, 56*) described a detailed study of transferrin, an iron-containing protein. Suffice it here to outline some of the basic principles. Since the activation enthalpy of a chemical exchange of water molecules or protons would differ in size or sign from the activation enthalpies expected for the different relaxation mechanisms the two limiting cases of Eq. (19), where T_1 would be controlled either by τ_{pr} or T_{1pr}, can in general be distinguished by a study of the temperature dependence of T_1. If T_{1pr} is found to control its origin may be further traced by variation of the magnetic field strength, and comparison of the observed field dependence with the behaviour expected from the dispersion terms in Eq. (20) and (21).

The major influence that a diamagnetic protein must have on the solvent protons with which it interacts is to increase τ_r from its value of ca. $3 \cdot 10^{-12}$ sec in the bulk water (*7*) to a value close to the tumbling time of the protein molecule. If the latter is assumed to be ca. $1 \cdot 10^{-7}$ sec for a spherical particle with the size of hemoglobin (*20*) one would expect from Eq. (20) a dispersion of $(T^d_{1pr})^{-1}$ centered around a proton resonance frequency of ca. 1 Mc, which corresponds to a magnetic field of 235 gauss. $(T^d_{1pr})^{-1}$ would then be much larger at very low fields than at the usual field strengths of several Kgauss. If T^d_{1pr} contains contributions from more than one type of hydration sites, this "nuclear magnetic relaxation dispersion" (*Koenig* and *Schillinger* (*55*)) may occur at different fields for the different contributions. For example the protein with $\tau_r \approx 1 \cdot 10^{-7}$ sec might also interact with loosely bound waters, for which τ_{pr} might be ca. $1 \cdot 10^{-8}$ sec, and hence short compared to τ_r. From Eq. (20) one would then expect a dispersion of T^d_{1pr} at a resonance frequency of ca. 10 Mc. The nuclear magnetic relaxation dispersion data could thus provide information on the relative contributions to the relaxation enhancement from the different hydration sites. The paramagnetic relaxation enhancement (Eq. (21)) can be analysed in a similar fashion through investigation of the dependence on the magnetic field strength.

The structural information derived from relaxation enhancement studies depends somewhat on the model chosen to describe the interaction of solvent protons with the protein molecules. For example even if the experiments indicated that the dispersion of T^d_{1pr} were essentially determined by the correlation time for rotational tumbling of the protein the tumbling of the hydration waters would not necessarily have to be restricted to that of the entire hydrated protein. Evidence was found that fast intramolecular tumbling about an axis fixed with respect to the surface of the hydrated species reduced the proton and O^{17} nuclear relaxation rates even in extremely stable aquocomplexes of Al^{3+} and other metal ions (*Connick* and *Wüthrich* (*21*)). The occurrence of similar intramolecular movements of the water molecules in hydrated proteins might greatly reduce the observed relaxation enhancement per hydration site, and hence affect the determination of the hydration number n in Eq. (19) (*Koenig* and *Schillinger* (*55*)). Additional complications might arise in the detailed interpretation of the data because protons could also be exchanged between the bulk water and certain amino acid residues of the protein.

B. Experiments with Hemoproteins

In the reported studies of hemoprotein solutions an enhancement of the proton relaxation in the bulk water was generally observed. From comparison with the diamagnetic proteins these effects were related mainly to interactions of the protons with the paramagnetic heme groups. Eq. (21) shows that these interactions would depend strongly on the access which solvent protons have to the heme groups. Starting with the work of *Davidson* and *Gold* (*23*) attempts were made to relate the observed relaxation enhancement to the position of the heme groups with respect to the protein surface (*Kon* and *Davidson* (*57*); *Wishnia* (*109*); *Lumry et al.* (*69*); *Fabry* and *Koenig* (*31*); *Scheeler* (*96*); *Mildvan et al.* (*81*)) and to possible exchange of protons or entire water molecules in and out of the sixth coordination site of the heme iron (*Fabry* and *Reich* (*32*); *Maričić et al.* (*72*); *Mildvan et al.* (*81, 82*); *Fabry et al.* (*33*)). Most of the earlier measurements were done at constant magnetic field and constant temperature. It was therefore in general not possible to distinguish between different relaxation mechanisms, and hence only qualitative data on the structure of the hemoprotein molecules were obtained. More recently the relaxation studies were extended to variable temperatures, and nuclear magnetic relaxation dispersion studies similar to those reported for transferrin were applied to solutions of myoglobin and hemoglobin [6]. From these very recent experiments a more complete analysis of the hydration of myoglobin and hemoglobin should emerge.

Because of the r^{-6}-dependence of the dipolar relaxation (Eq. 21), and the possibility of additional relaxation through contact interactions (Eq. 5), one might expect relatively large contributions to the proton relaxation in the bulk water from water or proton exchange in and out of the sixth coordination site of the heme iron (Fig. 3). Measurements of the exchange rates of protons (*Luz* and *Shulman* (*70*)) and entire water molecules (*Zeltmann* and *Morgan* (*119*); *Judkins* (*49*)) between the bulk water and the first coordination sphere of low molecular weight iron(III)-complexes indicate that these reactions might be sufficiently fast to be observed, provided that the heme iron is accessible in the proteins. From the different reports (*32, 72, 81, 82*) it appears that at present proton exchange from this position could not be excluded from consideration, and that it might actually be one of the dominant factors under certain conditions of pH and temperature.

Transitions between the high and low spin states of ferric hemoproteins (Table 1) greatly affect the relaxation enhancement, because T_{1e} and S in Eq. (21) would be greatly different for the two states (sections IV and VI). Even the presence of very small amounts of high spin ferric compounds should be detectable by relaxation enhancement experiments. For example *Kowalsky* (*63*) proposed that the dependence on temperature of the relaxation enhancement in cytochrome c solutions might be governed by a thermal equilibrium between high and low spin ferricytochrome c, where only very little high spin state would be present at any given moment.

Acknowledgements

I am greatly indepted to Prof. *R. Schwyzer* for his encouragement during the course of this work, and to Prof. *Hs. H. Günthard* for the hospitality extended to me at the Institute of Physical Chemistry of the Eidgenössische Technische Hochschule.

Much of the author's involvement with hemoproteins came during a two-year term of employment with Bell Telephone Laboratories in Murray Hill, N. J. The hospitality during this time in Dr. *R. G. Shulman's* department is greatfully acknowledged, and I wish to thank in particular Drs. *W. E. Blumberg, L. C. Snyder* and *T. Yamane* for stimulating discussions.

VIII. Appendix: New Experiments
(September 1970)

Since this review had been written several important new results of NMR studies in hemes and hemoproteins were reported. In the following these new data are briefly surveyed.

A. Low Spin Porphyrin-Iron(III) Complexes

Hill and *Morallee* (*121, 122*) recorded the proton NMR spectra of a series of bis-pyridinoiron (III) complexes. The compounds differed in the substituents at the ring positions 3 or 4 of the axial ligands, and the basicity of these ligands was thus varied from $pK_a = 1.9$ in 4-cyanopyridine to $pK_a = 9.2$ in 4-aminopyridine. In CD_3OD at ca. 200°K all these complexes were found to be in the low spin forms, and it was observed that the hyperfine shifts of the ring methyl resonances of protoporphyrin IX decrease linearly with increasing basicity of the axial ligands. For the mesoproton resonances the dependence on the axial ligands is much smaller and has the opposite sign. From this the authors concluded that the mesoproton resonances experience sizeable pseudocontact shifts so that the spin destities at the mesopositions would be considerably smaller than the upper limits given in Table 4. The main mechanism for spin delocalization would then appear to be a charge transfer from a ligand π-orbital to the iron, becoming less important as the electron density on the iron increases due to decrease in the basicity of the axial ligands. In a molecular orbital description the porphyrin π-orbital most strongly involved in this charge transfer would be the energetically highest-lying of the filled E_g-orbitals (see Section IV. E. 3). This orbital has nodes at the mesocarbon positions (*Longuet-Higgins et al.* (*68*)), and hence this charge transfer mechanism would explain that larger spin densities arise at the ring methyls than at the mesoprotons. A similar charge transfer mechanism was proposed by *R. G. Shulman* and *M. Karplus* (*122*) for the spin delocalisation in the low spin ferric heme compounds of Figs. 11—19.

B. Cytochrome c

New measurements of cytochrome *c* were made by *R. K. Gupta* and *A. G. Redfield* (*122, 123*), using a 100 MHz spectrometer converted to pulsed operation. In their experiments a pulse of radiofrequency radiation is applied at the resonance frequency of one of the hyperfine-shifted lines of ferricytochrome *c* (Fig. 19). The intensity of the pulse is fixed at a level just sufficient to saturate the resonance to which it is applied. The rest of the spectrum is then scanned by an observation pulse just after the saturating pulse to search for the changes induced by the latter. Since these changes are small they are recorded most easily by taking a difference spectrum with and without the saturating pulse. In this way changes can be observed even if they occur in the spectral region of the strongly overlapping polypeptide resonances (Fig. 8).

In mixed solutions of ferro- and ferri-cytochrome *c* cross saturation effects could be observed by this technique. For example when the methyl resonance at $+23.2$ ppm of ferricytochrome *c* (Fig. 19) was irradiated, saturation effects were also observed in the methyl resonance of ferrocytochrome *c* at $+3.3$ ppm (Fig. 27). This cross relaxation was shown to arise from an exchange of protein molecules, and hence also the saturated spins, between the ferrous and ferric oxidation states. The life-time in either oxidation state then has to be comparable to or shorter than the longitudinal spin relaxation time of the observed protons. Besides establishing an exchange of protein molecules between the two oxidation states this experiment shows that the resonances at $+23.2$ ppm in ferricytochrome c and at $+3.3$ ppm in ferrocytochrome c correspond to the same protons, which is in agreement with the resonance assignments presented in Section V. Similarly the assignment of the lines at -34.0 and -31.4 ppm (Fig. 19) to ring methyl groups of heme c (110) was confirmed, and the remaining two ring methyls in ferricytochrome c appear to be at -10.3 and -7.2 ppm. Two possible assignments are proposed for the methyl resonances at $+2.1$ and $+2.5$ ppm, which would correspond either to the methyl groups in the β-positions of the 2,4-substituents of heme c (Fig. 2) or to pseudocontact-shifted methyl resonances of aliphatic amino acid residues.

From these measurements the resonance positions in ferrocytochrome c of the protons corresponding to the hyperfine-shifted lines in ferricytochrome c are obtained. *Gupta* and *Redfield* (123) found that one of the ring methyl resonances of heme c is shifted upfield by ca. 1.5 ppm, indicating close proximity to the face of an aromatic amino acid residue in the reduced cytochrome c. The resonance positions in ferrocytochrome c are probably a better approximation of the "diamagnetic heme resonance positions" in ferricytochrome c than are the porphyrin-zinc (II) complexes (Fig. 9). Hence experiments of this type could conceivably contribute towards a more accurate determination of the spin density distribution in the heme groups.

Gupta and *Redfield* (123) studied also the binding of azide to ferricytochrome c. The three azidoferricytochrome c resonances at -17.3, -16.1, and -14.8 ppm (111) were found to come from the three ring methyls of heme c which are at -10.3, -34.0, and -7.2 ppm in ferricytochrome c. Values were obtained for the association and dissociation constants for the binding of azide ion to ferricytochrome c.

C. Hemoglobin

It has recently been possible by several types of NMR experiments to observe spectral changes attributable to subunitinteractions during ligand binding to tetrameric hemoglobins. *S. Ogawa* (122, 124) reported new measurements of hybrid model compounds (see Section VI. D. 2), where small effects of ligand binding on the heme resonances in the neighboring subunits were observed. *K. Winterhalter* and *K. Wüthrich* (124) studied some rather drastically modified tetrameric hemoglobin derivatives and found that the resonances of low spin ferric heme groups are not very sensitive to structural changes in the neighboring subunits. Hence it would seem that the spectral changes observed by *Ogawa* might conceivably correspond to quite extensive modifications of the protein conformation.

D. J. Patel (124) studied aqueous solutions of diamagnetic oxyhemoglobin and observed several resolved resonances between ca. − 11 and − 14 ppm. Since these protons are exchanged against deuterium in D_2O the unusual resonance positions were assigned to amide protons and to the indole-NH protons of the tryptophanyl residues. Since the positions of these lines changed upon deoxygenation it was concluded that they can be used to detect conformational changes during ligand binding to tetrameric hemoglobins. *M. A. Raftery (122, 124)* reported studies of hemoglobins which were modified by trifluoroalkylation at the cysteinyl residues 93 of the β-chains. Variations of the fluorine resonance positions were observed upon binding of various ligands, indicating that the fluorine nuclei are a useful probe for detecting subunitinteractions. All these different types of measurements promise to be useful for further studies of the mechanism of ligand binding to hemoglobin.

IX. References

1. *Abragam, A.:* The Principles of Nuclear Magnetism. London: Oxford University Press 1961.
2. − *Pryce, M. H. L.:* Proc. Roy. Soc. (London) *A 206*, 173 (1951).
3. *Antonini, E.:* Physiol. Rev. *45*, 123 (1965).
4. − Science *158*, 1417 (1967).
5. *Becker, E. D., Bradley, R. B.:* J. Chem. Phys. *31*, 1413 (1959).
6. − − *Watson, C. J.:* J. Am. Chem. Soc. *83*, 3743 (1961).
7. *Beetlestone, J., George, P.:* Biochemistry *3*, 707 (1964).
8. *Bersohn, R.:* J. Chem. Phys. *24*, 1066 (1956).
9. *Bloembergen, N.:* J. Chem. Phys. *27*, 595 (1957).
10. *Blumberg, W. E., Peisach, J.:* Paper presented at the 1 Simposio Interamericano sobre Hemoglobinas, Caracas, Venezuela, December 3—5, 1969.
11. *Bolton, J. R., Carrington, A., McLachlan, A. D.:* Mol. Phys. *5*, 31 (1962).
12. *Bohr, C., Hasselbalch, K., Krogh. A.:* Scand. Arch. Physiol. *16*, 402 (1904).
13. *Bolton, W., Cox, J. M., Perutz, M. F.:* J. Mol. Biol. *33*, 283 (1968).
14. *Bovey, F. A.:* Nuclear Magnetic Resonance Spectroscopy. New York: Academic Press 1969.
15. *Bradshaw, R. A., Gurd, F. R. N.:* J. Biol. Chem. *244*, 2167 (1969).
16. *Carrington, A., McLachlan, A. D.:* Introduction to Magnetic Resonance. New York: Harper and Row 1967.
17. *Caughey, W. S., Koski, W. S.:* Biochemistry *1*, 923 (1962).
18. − *Ibers, P. K.:* J. Org. Chem. *28*, 269 (1963).
19. − *Johnson, L. F., Wüthrich, K., Shulman, R. G.:* Proc. of the 3rd International Conference on Magnetic Resonance in Biological Systems, Warrenton Va., October 14—18, 1968.
19a. − *Johnson, L. F.:* Chem. Comm. *1969*, 1362.
20. *Cohn, E. J., Edsall, J. T.:* Proteins, Amino Acids and Peptides, p. 557. New York: Reinhold Publishing Company 1943.

21. *Connick, R. E., Wüthrich, K.:* J. Chem. Phys. *51*, 4506 (1969).
22. *Dayhoff, M. O.,* ed.: Atlas of Protein Sequence and Structure. Silver Spring Md: Nat. Biomed. Res. Found. 1969.
23. *Davidson, N., Gold, R.:* Biochim. Biophys. Acta *26*, 370 (1957).
24. *Dickerson, R. E.:* In: The Proteins, vol. 2 (2nd ed.) p. 603; ed. by *H. Neurath.* New York: Academic Press 1964.
25. — *Kopka, M. L., Weinzierl, J. E., Varnum, J. C., Eisenberg, D., Margoliash, E.:* J. Biol. Chem. *242*, 3015 (1967).
26. — — — — — In: Structure and Function of Cytochromes, p. 225; ed. by *Okunuki, K., Kamen, M. D., Sekuzu, I.* Baltimore Md: University Park Press 1968.
27. *Eaton, D. R., Josey, A. D., Phillips, W. D., Benson, R. E.:* J. Chem Phys. *37*, 347, (1962).
28. — — *Benson, R. E., Phillips, W. D., Cairns, T. L.:* J. Am. Chem. Soc. *84*, 4100 (1962).
29. *Edmundson, A. B.:* Nature *205*, 883 (1965).
30. *Esposito, J. N., Lloyd, J. E., Kenney, M. E.:* Inorg. Chem. *5*, 1979 (1966).
31. *Fabry, T. L., Koenig, S. H.:* In: Hemes and Hemoproteins, p. 241; ed. by *Chance, B., Estabrook, R., Yonetani, T.* New York: Academic Press 1966.
32. — *Reich, H. A.:* Biochem. Biophys. Res. Commun. 22, 700 (1966).
33. — *Kim, J., Koenig, S. H., Schillinger, W. E.:* In the Proceedings of the 4th Johnson Research Foundation Colloquium, April 19—21, 1969. New York: Academic Press (in print).
34. *George, P., Schejter, A.:* J. Biol. Chem. *239*, 1504 (1964).
35. *Gibson, J. F., Ingram, D. J. E.:* Nature *180*, 29 (1957).
36. *Griffith, J. S.:* Nature *180*, 30 (1957).
37. — The Theory of Transition-Metal Ions. Cambridge: Cambridge University Press 1964.
38. — Biopolymers Symp. *1*, 35 (1964).
39. *Guidotti, G.:* Paper presented at the 1 Simposio Interamericano sobre Hemoglobinas, Caracas, Venezuela, December 3—5, 1969.
40. *Helcké, G. A., Ingram, D. J. E., Slade, E. F.:* Proc. Roy. Soc. (London) B *169*, 275 (1968).
41. *Hoard, J. L.:* In: Structural Chemistry and Molecular Biology, p. 573; ed. by *Rich, A., Davidson, N.* New York: Freeman 1968.
42. *Hogness, T. R., Zscheile, F. P., Sidwell, A. E., Barron, E. S. G.:* J. Biol. Chem. *118*, 1 (1937).
43. *Horecker, B. L., Kornberg, A.:* J. Biol. Chem. *165*, 11 (1946).
44. *Horrocks, W. D., Taylor, R. C., LaMar, G. N.:* J. Am. Chem. Soc. *86*, 3031 (1964).
45. *Hugli, T. E., Gurd, F. R. N.:* J. Biol. Chem. *245*, 1930, 1939 (1970).
46. *Jesson, J. P.:* J. Chem. Phys. *47*, 579 (1967).
47. — J. Chem. Phys. *47*, 582 (1967).
48. *Johnson, C. E., Bovey, F. A.:* J. Chem. Phys. *29*, 1012 (1958).
49. *Judkins, M. R.:* Ph. D. Thesis. University of California, Lawrence Radiation Laboratory Report Nr. UCRL-17561. Berkeley, Calif. 1967.
50. *Keilin, D.:* Proc. Roy. Soc. (London) B *98*, 312 (1925).
51. *Kendrew, J. C.:* Sci. Am. *205* (6), 96 (1961).
52. — Science *139*, 1259 (1963).
53. — *Watson, H. C , Strandberg, B. E., Dickerson, R. E., Phillips, D. C., Shore, V. C.:* Nature *190*, 666 (1961).

54. *Kilmartin, J. V., Rossi-Bernardi, L.:* Nature *222*, 1243 (1969).
55. *Koenig, S. H., Schillinger, W. E.:* J. Biol. Chem. *244*, 3283 (1969).
56. — — J. Biol. Chem. *244*, 6520 (1969).
57. *Kon, H., Davidson, N.:* J. Mol. Biol. *1*, 190 (1959).
58. *Koshland, D. E., Némethy, G., Filmer, D.:* Biochemistry *5*, 365 (1966).
59. *Kotani, M.:* Suppl. Progr. Theoret. Phys. *17*, 4 (1961).
60. — Advan. Chem. Phys. *7*, 159 (1964).
61. *Kowalsky, A.:* J. Biol. Chem. *237*, 1807 (1962).
62. — Biochemistry *4*, 2382 (1965).
63. — Federation Proc. *28*, 603 (1969).
64. *Kurland, R. J., Davis, D. G., Ho, C.:* J. Am. Chem. Soc. *90*, 2700 (1968).
65. *LaMar, G. N.:* J. Chem. Phys. *43*, 1085 (1965).
66. — *Horrocks, W. D., Allen, L. C.:* J. Chem. Phys. *41*, 2126 (1964).
67. *Lang, G., Marshall, W.:* Proc. Phys. Soc. (London) *87*, 3 (1966).
68. *Longuet-Higgins, H. C., Rector, C. W., Platt, J. R.:* J. Chem. Phys. *18*, 1174 (1950).
69. *Lumry, R., Matsumiya, H., Bovey, F. A., Kowalsky, A.:* J. Phys. Chem. *65*, 837 (1961).
70. *Luz, Z., Shulman, R. G.:* J. Chem. Phys. *43*, 3750 (1965).
71. *Margoliash, E., Schejter, A.:* Advan. Protein Chem. *21*, 113 (1966).
71a. *Margoliash, E.:* In: Hemes and Hemoproteins, p. 371; ed. by Chance, B., Estabrook, R., Yonetani, T. New York: Academic Press 1966.
72. *Maričić, S., Ravilly, A., Mildvan, A. S.:* In: Hemes and Hemoproteins, p. 157; ed. by *Chance, B., Estabrook, R., Yonetani, T.* New York: Academic Press 1966.
73. *McConnell, H. M.:* J. Chem. Phys. *24*, 764 (1956).
74. — J. Chem. Phys. *28*, 430 (1958).
75. — *Chesnut, D. B.:* J. Chem. Phys. *28*, 107 (1958).
76. — *Robertson, R. E.:* J. Chem. Phys. *29*, 1361 (1958).
77. *McDonald, C. C., Phillips, W. D.:* J. Am. Chem. Soc. *89*, 6332 (1967).
78. — — J. Am. Chem. Soc. *91*, 1513 (1969).
79. — — *Vinogradov, S. N.:* Biochem. Biophys. Res. Commun. *36*, 442 (1969).
80. *Mc Lachlan, A. D., Muirhead, H., Perutz, M. F.:* J. Mol. Biol. (in print).
81. *Mildvan, A. S., Rumen, N. M., Chance, B.:* Federation Proc. *27*, 525 (1968).
82. — — — In the Proceedings of the 4th Johnson Research Foundation Colloquium, April 19—21, 1969. New York: Academic Press (in print).
83. *Nelson, F. A., Weaver, H. E.:* Science *146*, 223 (1964).
84. *Ogawa, S., McConnell, H. M.:* Proc. Natl. Acad. Sci. U.S. *58*, 19 (1967).
85. — — *Horwitz, A.:* Proc. Natl. Acad. Sci. U.S. *61*, 401 (1968).
86. *Ohno, K., Tanabe, Y., Sasaki, F.:* Theoret. Chim. Acta *1*, 378 (1963).
87. *Pauling, L.:* J. Chem. Phys. *4*, 673 (1936).
88. — *Coryell, C. D.:* Proc. Natl. Acad. Sci. U.S. *22*, 210 (1936).
89. *Perutz, M. F.:* J. Mol. Biol. *13*, 646 (1965).
90. — *Muirhead, H., Cox, J. M., Goaman, L. C. G.:* Nature *219*, 131 (1968).
91. — — *Mazzarella, L., Crowther, R. A., Greer, J., Kilmartin, J. V.:* Nature *222*, 1240 (1969).
92. *Pfeifer, H.:* Biochim. Biophys. Acta *66*, 434 (1963).
93. *Pullman, B., Spanjaard, C., Berthier, G.:* Proc. Natl. Acad. Sci. U.S. *46*, 1011 (1960).
94. *Roberts, G. C. K., Jardetzky, O.:* Advan. Protein Chem. *24*, 447 (1970).
95. *Rossi-Fanelli, A., Antonini, E., Caputo, A.:* Advan. Protein Chem. *19*, 73 (1964).
95a. *Salmeen, I., Palmer, G.:* J. Chem. Phys. *48*, 2049 (1968).

96. *Scheler, W.:* Biochim. Biophys. Acta *66,* 424 (1963).
97. *Sheard, B.:* Paper presented at the 1 Simposio Interamericano sobre Hemoglobinas, Caracas, Venezuela, December 3—5, 1969.
98. *Shulman, R. G., Wüthrich, K., Yamane, T., Antonini, E., Brunori, M.:* Proc. Natl. Acad. Sci. U.S. *63,* 623 (1969).
99. — *Ogawa, S., Wüthrich, K., Yamane, T., Peisach, J., Blumberg, W. E.:* Science *165,* 251 (1969).
100. — *Wüthrich, K., Peisach, J.:* In the Proceedings of the 4th Johnson Research Foundation Colloquium, April 19—21, 1969. New York: Academic Press (in print).
101. — — *Yamane, T., Patel, D., Blumberg, W. E.:* J. Mol. Biol. (in print).
102. *Solomon, I.:* Phys. Rev. *99,* 559 (1955).
103. *Sternlicht, H., Wilson, D.:* Biochemistry *6,* 2881 (1967).
104. *Swift, T. J., Connick, R. E.:* J. Chem. Phys. *37,* 307 (1962).
105. *Theorell, H.:* J. Am. Chem. Soc. *63,* 1820 (1941).
106. *Weissbluth, M.:* In: Structure and Bonding, vol. 2, p. 1, Berlin–Heidelberg–New York: Springer 1967.
107. *Weissman, S. I.:* J. Chem. Phys. *25,* 890 (1956).
108. *Winterhalter, K. H., Deranleau, D. A.:* Biochemistry *6,* 3136 (1967).
109. *Wishnia, A.:* J. Chem. Phys. *32,* 871 (1960).
110. *Wüthrich, K.:* Proc. Natl. Acad. Sci. U.S. *63,* 1071 (1969).
111. — In the Proceedings of the 4th Johnson Research Foundation Colloquium, April 19—21, 1969. New York: Academic Press (in print).
112. — *Shulman, R. G., Peisach, S.:* Proc. Natl. Acad. Sci. U.S. *60,* 373 (1968).
113. — — *Yamane, T.:* Proc. Natl. Acad. Sci. U.S. *61,* 1199 (1968).
114. — — *Wyluda, B. J., Caughey, W. S.:* Proc. Natl. Acad. Sci. U.S. *62,* 636 (1969).
115. — — *Yamane, T., Wyluda, B. J., Hugli, T. E., Gurd, F. R. N.:* J. Biol. Chem. *245,* 1947 (1970).
116. — — — *Blumberg, W. E.:* to be published.
117. *Wyman, J.:* Advan. Protein Chem. *19,* 223 (1964).
118. *Yamane, T., Wüthrich, K., Shulman, R. G., Ogawa, S.:* J. Mol. Biol. *49,* 197 (1970).
119. *Zeltmann, A. H., Morgan, L. O.:* J. Phys. Chem. *70,* 2807 (1966).
120. *Zerner, M., Gouterman, M., Kobayashi, H.:* Theoret. Chim. Acta *6,* 363 (1966).
121. *Hill, H. A. O., Morallee, K. G.:* Chem. Comm. *1970,* 266.
122. Presented at the Fourth International Conference on Magnetic Resonance in Biological Systems, Oxford, England, August 26—September 2, 1970.
123. Personal communication. Papers are in press for Science, Biochem. Biophys. Res. Comm., and the Proceedings of the Wenner-Gren Symposium.
124. Presented at the 8th International Congress of Biochemistry, Interlaken, Switzerland, September 3—9, 1970.

Received February 2, 1970

The Chemical Nature and Reactivity of Cytochrome P-450

Dr. H. A. O. Hill, Dr. A. Röder*, and Dr. R. J. P. Williams

Inorganic Chemistry Laboratory, Oxford, England

Table of Contents

* Present address: Chemisches Institut, Lehrstuhl für Organische Chemie, BRD, 74 Tübingen, Wilhelmstr. 33, Germany.

H. A. O. Hill, A. Röder, and R. J. P. Williams

I. Introduction

P-450 is the name given to a protoheme-containing protein which, upon reduction in the presence of carbon monoxide, gives an intense absorption band, the Soret band, at the unusually long wavelength for these complexes of about 450 nm. It was first reported (*1*) in rat liver microsomes but since then a large family of related proteins has been discovered, not only in microsomes and mitochondria but also in bacteria (*1—8*). Even in individual biological systems it now seems clear that several different P-450 enzymes can be generated. The interest in these proteins lies in their oxidative activity associated with detoxification reactions in membranes. The substrates which can be oxidised include long-chain aliphatic compounds, steroids, cyclohydrocarbons, condensed aromatics, anilines, phenols and barbiturates, illustrating the low specificity of the reaction. The oxidation takes the form of the insertion of hydroxyl groups and it can be used in demethylation. It is the purpose of this review to draw attention to the *unusual state of the metal-porphyrin complex in the protein* and to suggest the importance this may have in catalysis.

The enzymes have two other substrates in that they require molecular oxygen and a source of reducing equivalents (*4*). As they convert one atom of the oxygen molecule to water and the other to the hydroxyl group inserted into the substrate, they are called mixed function oxidases. The reaction can be written:

$$SH_2 + O_2 + H_2 \longrightarrow SHOH + H_2O$$

The source of the reducing equivalents is a short electron-transfer chain which starts from reduced nicotinamide, NADH or NADPH. Fig. 1 shows the generally accepted scheme of the reaction (*4*).

In this review we shall concentrate upon the properties of P-450 itself, making only casual reference to the other protein components of Fig. 1 and NADH. However, as the oxygen is effectively acting in a

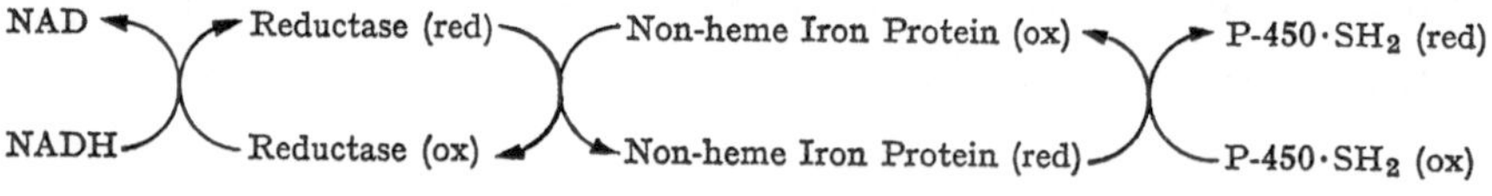

$$P\text{-}450 \cdot SH_2 \text{ (red)} + O_2 \longrightarrow P\text{-}450 \text{ (ox)} + SHOH + H_2O$$

Fig. 1. The pathway of oxidising and reducing equivalents in the microsomal redox chain, (red) is reduced and (ox) is oxidised.

124

four-electron process the problem of introducing these electrons must be considered. This is not a new problem in biological systems for it occurs in the terminal oxidases, e.g. cytochrome oxidase, and in many plant oxidases, e.g. laccases and ascorbic acid oxidase. In all these cases oxygen is reduced directly to water and there is no known intermediate formation of hydrogen peroxide. Like P-450 then these enzymes have a large number of redox centres coupled together and it could well be that the chain of enzymes associated with P-450 is deliberately introduced so as to avoid hydrogen peroxide formation. One of the difficulties of preparing model systems for the study of P-450 is the inability of chemists to construct such multi-electron sources and sinks, see section V.

We turn now to a discussion of the properties of P-450 itself.

II. Chemical Properties of P-450

a) The Nature of the Protein

Early attempts at preparing P-450 from membranes were thwarted by the loss of absorption in the 450 nm region (CO complex) and the appearance in its place of absorption at 420 nm (1, 2, 5, 7). The P-420 pigment preserves all the characteristics of a heme-protein and the only heme present is protoheme. The spectra of the P-420 complex are given in Table 1 as they provide a basis for the discussion of P-450. The P-420 spectra are in all respects those of normal hemochromogens i.e. simple six-coordinate Fe(II) and Fe(III) hemo-proteins such as cytochrome b.

Table 1. *Optical spectra of P-420 at pH 7.5*

	$\lambda(\gamma)$ nm	ε x 10^{-3}	$\lambda(\beta)$ nm	ε x 10^{-3}	$\lambda(\alpha)$ nm	ε x 10^{-3}
Oxidised Fe(III)	414	124	535	11	—	—
Reduced Fe(II)	427	149	530	13	559	24
Reduced Fe(II).CO	421	213	538	17	565	16

Various reagents bring about the change from P-450 to P-420 including proteases, phospholipases, alcohols, ketones, detergents and sulphydryl reagents (e.g. parachloromercury benzoate and CH_3I) (10—18). Many of these reactions are reversible (14). An extended study (15, 16) showed that both for urea and phenol derivatives the conversion

125

to P-420 increased as the partition coefficient of the added reagent, between octanol and water, increased. Such observations led to the suggestions (*17, 18*) that: (1). Hydrophobic interactions are very important in the structure of P-450, and that (2) the release of the P-450 from the lipid membrane materially disrupted its integrity. The separation of very clean P-450 from several sources now makes (2) an unnecessary postulate. It remains the case that P-450 can interact non-specifically with a wide variety of hydrophobic compounds as is clear from the nature of its substrates.

The isolation of a stable P-450 from membranes by *Horie* and *Kinoshita* (*22*), *Appleby* (*3*), and *Gunsalus et al.* (*6*) has revealed the nature of this pigment. The optical absorption characteristics are given in Table 2 for comparison with Table 1. The isolated P-450 shows all of the properties of the membrane bound compound including the sensitivity to hydrophobic reagents and mercurials.

Table 2. *The spectra of P-450 in oxidised and reduced states*

		Spectra of P-450			Hb[a] Corresponding	
		γ	β	α	Compound	
a) Reduced	—	412 (low)	555	?	432	555
	CO	450	558	none	420	541 568
	EtNC low *p*H	429	530	<557	428	530 <560
	high *p*H	454	551	>(580)		
	CN⁻	432	532	560	428	534 <561
	NO		550 flat			
N.B. P-420		427	530	<559?		
	CO	421	538	565		
b) Oxidised	—	416	535 570	650 low	406	500 630
	ETNC	430	550			
	CN⁻	436	553		420	540 (575)
	N₃⁻				418	540 (575)
	NO	432	543	575		
	RS⁻				545	578
N.B. P-420		414	535	—		
Cyt c (reduced)			530	<565		
Cyt b (reduced)			530	<560		

[a] Hb is haemoglobin.

Much of the earlier confusion as to the nature of microsomal P-450 was removed when it was discovered that the biosynthesis of the "different" P-450 proteins was enhanced by the known substrates (8, 9). In particular P-450 induced by 3-methyl cholanthrene has a high extinction coefficient at around 390—400 nm as compared with the P-450 induced by 3,4-benzpyrene when the peak of the oxidised form is at 421 nm. All in all it now appears that two extremes of spectral types of P-450 exist (19—24): type (I) is induced by 17-hydroxyprogesterone, deoxycortisone, benzpyrenè, cyclohexane or camphor and has high absorption in the 390 region. The absorption moves to shorter wavelength on addition of substrate but after hydroxylation returns to its previous position: type (II) is induced by other substrates or by combination with aniline, imidazoles and other bases and corresponds to a shift of the Soret band to longer wavelengths, increase at 420 nm and decrease at 390 nm. These spectra are little altered at low temperatures, even at that of liquid nitrogen.

b) Complexes of P-450

There is surprisingly little evidence in the literature about the binding of anions to oxidised P-450. Perhaps F^-, OH^-, N_3^-, etc. do not bind or have a very low affinity. The redox potential of P-450 is well below 0.0 volts and these properties may well indicate that the heme is already bound by an anion (e.g. RS^- see later). The majority of studies of complexes deal therefore with neutral, mostly unsaturated, ligands.

c) Carbon Monoxide, Base, and Cyanide Complexes

The reaction of reduced P-450 with carbon monoxide is the basis of its identification (1—8). The spectrum is however unusual in respects other than in the Soret region as we shall discuss in this section. The affinity of the reduced P-450 for carbon monoxide is rather low compared with that of hemeproteins such as myoglobin. The compound is photo-labile like other heme.CO complexes. P-450 binds cyanide in both the reduced and oxidised states. The binding is again very weak compared with that of myoglobin and perhaps this is due to the very hydrophobic nature of the environment of the heme. In most respects the cyanides appear to be quite normal although the Soret bands are at somewhat longer wavelength than expected. P-450 also binds to amines and *Jefcoate* and *Gaylor* (24—26) have shown that these amine complexes have absorption spectra with long-wavelength Soret bands at around 450 nm. P-420 forms a parallel series of complexes to those of P-450 but all of them appear to be quite normal as judged by their spectra.

d) Isonitrile Complexes (*13—15*)

The isonitrile complexes of P-450 have been studied in great detail because they show two curious properties for hemo-proteins. Firstly both Fe(II) and Fe(III) complexes are formed and secondly the Fe(II) complex changes its character with pH. The Fe(II) complex at low pH has absorption maxima at 429, 530 and 557 nm, typical of other heme isonitriles, while at high pH it has absorption bands at 454, 551 and no (or very weak) absorption at 580 nm. The second spectrum is very like that of the P-450 CO complex. An acid dissociation constant for the switch between the two forms has been calculated for different isonitriles and the pK_a falls as follows

methyl-$>$ ethyl-$>$ t-butyl-$>$ phenylisonitrile.

Clearly the isonitrile complexes are involved in an equilibrium which is similar to the virtually irreversible shift from P-450 to P-420 in the carbon monoxide complexes.

The Fe(III) isonitrile has absorption maxima at 433, 553 and 590 nm which are again long-wavelength bands for low-spin Fe(III) complexes.

e) Summary of Chemistry

The over-riding impression from these studies is that P-450 is *a protoheme protein of unusual properties* (*27, 28*). The heme lies in a hydrophobic pocket which is open to a wide variety of organic molecules many of which are very large but it is not open to anions. The neighbourhood of the iron may well contain a sulphydryl group and this reduces the interaction of ligand anions at the metal centre. The type of heme complex which is formed depends upon pH and minor conformational changes in the protein suggesting that there is more than one set of potential ligands for the heme iron within the hydrophobic pocket in which it sits. The protoheme group does not behave as expected in many of its physical properties and it is to these that we shall now turn.

III. Physical Properties of P-450

a) Electron Spin Resonance of P-450

The observation of Mason and his colleagues of an electron paramagnetic resonance signal in microsomes, $g = 1.91$, 2.25, 2.40, see Fig. 2, was made before it was known that this was a heme-iron signal (*29*). For a while the signal was also called 'Fe$_x$' but this is now known to be P-450. The signals were later shown to be induced by those 'Fe$_x$' substrates which induced P-450 as shown by spectroscopic measurements. The iron could only be low-spin iron(III), i.e. the oxidised state of the protein. *Mason (4)* drew attention to the possibility that the signal could arise from the association of heme iron with thiols but he did not test this hypothesis beyond noting that the activity of the enzyme was affected by mercury reagents known to react with various sulphur-groups.

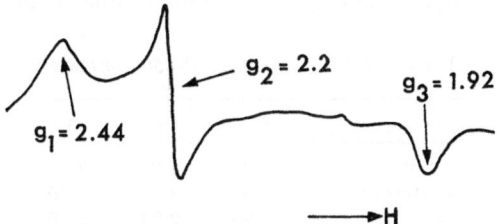

Fig. 2. The *g*-values in the electron spin resonance spectrum of P-450, oxidised

The EPR signals of P-450 are affected by *detergents, lowering of the pH*, and by *reagents for sulphydryl groups*, all of which can convert the low-spin Fe(III) to a high-spin form having a signal at $g = 6.1$ (*30, 31*). Again conversion to P-420 does not appear to alter the low-spin Fe(III) signals grossly and after treatment of the P-420 with sulphydryl reagents it too gives rise to a high-spin signal at $g = 6.1$.

An immediate working hypothesis from these results is that the heme-iron in P-450 and P-420 is bound in an unusual chemical form which gives rise to the rather curious low-spin signals — they are different from those in most other haem proteins, Table 3. Moreover these signals

Table 3. *g-values of low-spin hemoproteins*

	g_1	g_2	g_3	Ref.
Mb.OH	2.61	2.19	1.82	(*34*)
Mb.N$_3$	2.8	2.25	1.75	(*34*)
Hb.OH	2.6	2.3	1.7	(*35*)
Hb.N$_3$	2.82	2.2	1.70	(*35*)
Peroxidase.OH	2.86	2.12	1.67	(*41*)
Cytochrome *c* peroxidase	2.7	2.2	1.83	(*42*)
Cytochrome b_5	3.03	2.23	1.93	(*43*)
Catalase.N$_3$	2.80	2.18	1.74	(*36*)
	(2.53	2.2	1.85)	(*36*)
Catalase.CN	2.84	2.25	1.66	
	2.56	2.31	1.81	(*36*)
	(2.43	2.17	1.895)	

Mb is myoglobin and Hb is hemoglobin.

are not very sensitive to change in environment of the heme as shown by their persistence on switching from P-450 to P-420. This is confirmed too by the g-values of the Fe(III) isonitrile complexes (37): methylisonitrile 1.87, 2.25, 2.43; tertiary-butylisonitrile, 1.88, 2.2, 2.43; and phenyl-isonitrile, 1.89, 2.2 and 2.40. These values are very little different from those of P-450 itself and in fact *Estabrook et al.* (33) showed that only slightly different EPR shifts could be produced in the presence of substrates such as 17-hydroxyprogesterone and aniline, Table 4.

Table 4. *EPR g-values for P-450 complexes*

	g_1	g_2	g_3
P-450	1.91	2.25	2.41
+ steroids	1.93	2.24	2.40
+ aniline	1.90	2.24	2.45

The division of the substrates amongst those which increase the separation between the g values, e.g. aniline, and those which decrease the separation e.g. steroids is the same as the Type I and Type II division (32) of substrates based on absorption shifts (see above).

b) Model Complexes: EPR Spectra

The first suggestion that the EPR signals could be due to heme iron associated with sulphur ligands is due to Mason (37, 38), basing his conclusions on the general chemistry of the system (4). In a very different set of complexes *Röder* and *Bayer* (39) observed that cysteine and its derivatives will interact with a high-spin iron(III) trisethanolamine complex to give a set of EPR g-values not too different from those of P-450 and they too were led toward the conclusion that iron and sulphur were associated in P-450.

Subsequently two groups of workers have made heme model complexes specifically designed to show that the interaction of heme iron of P-450 could well be with a thiol group. *Bayer, Hill, Röder,* and *Williams* (45) prepared a series of Fe(III) hemoglobin and myoglobin thiol complexes, Table 5, which gave rise to the same pattern of EPR g-values as those seen in P-450 cytochromes. *Gaylor* and *Jefcoate* (40) made a similar study of the isopropyl-thiol of Fe(III) myoglobin and showed that it too gave the characteristic g values of P-450. It has been observed by *Hill, Röder* and *Williams* (46) that only certain thiols could make contact with the iron atom of myoglobin indicating that there is considerable steric hindrance at the iron. These observations will be discussed later as they have a bearing on oxidation.

Table 5. *EPR g values of metmyoglobin (Mb) and met-hemoglobin (Hb) complexes with sulphur ligands (45)*

	g_1	g_2	g_3
Mb H₂S	2.4	2.3	1.91
CH₃SH	2.38	2.25	1.94
C₂H₅SH	2.44	2.25	1.92
CH₃OOCCH₂SH	2.46	2.25	1.91
Hb H₂S	2.46	2.25	1.92
CH₃SH	(2.33	2.24	1.95)
	(2.44	2.24	1.93)
P-450	2.41	2.26	1.91

While stressing these parallels between EPR g-values in models containing sulphide links to iron and in P-450 we have to note that such g-values can be produced in other ways. *Röder (47)* reported an Fe(III)-cyanide-trisethanolamine complex with g-values of 2.41, 2.17 and 1.92 and Table 6 lists several examples of hemo-proteins in which a similar pattern of g-values has been seen. In the case of the β-chains of hemo-globins it could be that a rearrangement of the protein has permitted contact between the cysteine and the iron but this does not seem likely in many of the other proteins. At the present time it would seem unwise to draw up simple correlations between the nature of the chemical binding of the heme and the g-values. Notwith-standing these objections the g-values of P-450 and the models together with the extensive chemical and spectroscopic data strongly implicate RS⁻ binding to heme Fe(III) in P-450. For this reason we shall review here a more extensive study of heme Fe(III)RS⁻ complexes.

Röder and *Bayer (44)* reported the further study of a series of thiol/Fe(III) heme/base complexes, varying both the thiol and the base, the latter being either a pyridine or an aliphatic base. They showed that there was a correlation between the acid dissociation constants pK(RSH) or pK(base) and the g_1 and g_0 values, Fig. 3. This could well imply that,

Table 6. *Abnormal g-values in heme-systems*

	g_1	g_2	g_3	Ref.
Denatured Hb	1.9	2.2	2.4	(48)
Denatured β-chains of Hb	1.9	2.2	2.4	(49)
Lamprey HbOH	1.92		2.45	(50)
Hagfish HbOH	1.92			(50)
Glycera₃ Hb	1.95	2.21	2.33	(50)

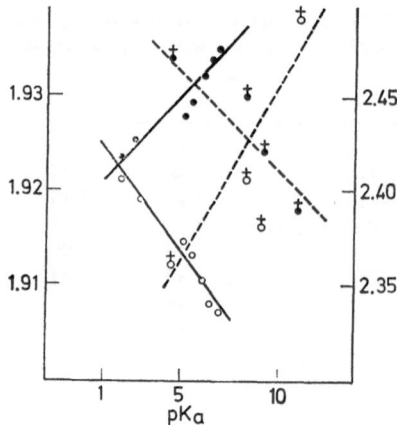

Fig. 3. The relationship between the position of the g-values in the EPR signal of P-450 model complexes and the pK_a values of the ligands to the iron. The different symbols in the figure refer to different series of ligands — see ref. (44)

at least in some instances, the g-value splitting is directly related to the σ-chemical bond-strength from the axial ligands to the iron. The stronger the donation of charge to the iron the nearer the g-values approach to 2.0. It was also observed that remote ionisations in the ligands, e.g. of the β-substituent of a thiol, could produce small g-value shifts, and that stericly-hindered ligands produced rather widely separated g-values.

A further correlation can be found between the g-value shifts in thiol-Fe(III)heme-base complexes varying the base for a constant thiol, and the NMR shifts on the periphal-CH_3 of the porphyrin for the same series of bases, Fig. 4. This correlation can be taken to confirm that the g-value shifts are mainly dependent upon electronic structure of the iron and not just minor geometric changes, given of course that electronic and geometric changes are intimately interwoven.

A comparison between this work and that on P-450 shows that the changes of the EPR signals have common features:

1. Unsaturated ligands such as isonitriles have a similar effect on P-450 to that of pyridines in models, e.g. g (i.e. $g < 2$) for methyliso-nitrile > phenylisonitrile.

2. Sterically hindered groups give less widely separated g-values e.g. t-butylisonitrile < methylisonitrile.

3. Type(II) substrates of P-450 which are better σ-donors than Type(I) substrates move the g values further apart, compare the effect of different amines in the models, Fig. 4.

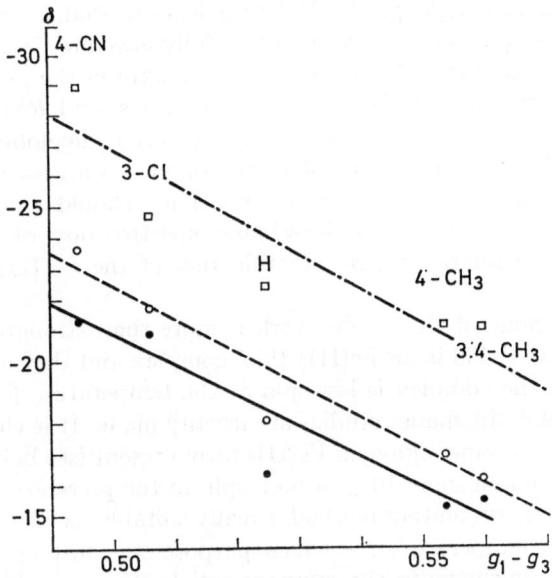

Fig. 4. Correlation between the shifts, delta (NMR), of the peripheral methyl protons (2 □, 1 O, 1 ●*b*) in low spin protoheme-Fe(III)-bis-pyridine (substituted) — complexes, (*Hill and Morallee*) and the splitting of the *g*-values g_1—g_3 (EPR) of the corresponding pyridine-methylthioglycollate-protoheme-complexes

An extreme effect of a coordinating ligand or an adjustment of the protein in P-450 could lead to a breaking of the Fe-RS⁻ bond and this could explain the observation of high-spin signals. Type(I) substrates do appear to reduce the intensity of the low-spin signals and although they do not bind the iron they could well weaken its binding to the protein. *Gaylor* and *Jefcoate (26)* have reported high-spin EPR signals of methylcholate induced P-450. These high-spin signals are of interest for they are asymmetric around $g = 6.0$. Such asymmetric signals are not normally seen in hemoproteins but have been seen in catalase, Table 7. Catalase

Table 7. *EPR g-values of some high-spin hemeproteins*

	g_1	g_2	g_3	Ref.
Metmyoglobin	5.9	—	2.0	*(34)*
Catalase	6.6	5.4	2.0	*(36)*
+F⁻	6.5	5.5	2.0	
+N₃⁻	6.7	5.2	2.0	
P-450	6.6	(5.3)	2.0	*(26)*

is curious amongst high-spin Fe(III) complexes in that it has very weak axial ligands — the cation never becomes fully low-spin. It could be that the high-spin heme of P-450 is also bound weakly in the protein.

In collaboration with *Lang, Winter, Williams,* and *Röder* have also studied the Mössbauer spectrum of the Fe(III) hemoglobin thiol-complexes. The Mössbauer spectra of these complexes are rather different from those of all other low-spin complexes. Thus, should it prove possible to substitute ^{57}Fe in P-450, Mössbauer spectroscopy of the protein could lead to a relatively safe identification of the Fe(III) heme/thiol link.

Although none of this model work is more than strongly suggestive it is likely that P-450 is an Fe(III) thiol complex and that a fairly large proportion of the complex is low-spin at the temperature (—180 °C) at which EPR and Mössbauer studies are usually made. It is clear however that there is also some high-spin Fe(III) form present (see below) but it is difficult to estimate some 10% of high-spin in the presence of 90% low-spin by EPR. Again neither method is really suitable for testing the spin-state at room temperature for which purpose we must use absorption spectroscopy. As shown in the previous article the amount of high-spin Fe(III) in the mixed spin equilibrium can be easily estimated by using the correlation between the magnetic and optical properties (*52*).

An entirely distinct series of model complexes has been carried out in order to show that metal porphyrins will actually bind to the type of substrate with which P-450 interacts. *Hill, Macfarlane, Mann,* and *Williams (51)* have studied molecular complex formation between such molecules as quinones and sterols and several metal porphyrins. The complexes between some of the porphyrins and sterols are remarkable strong. At the same time they have devised NMR methods for the elucidation of the structures of these complexes.

c) Absorption Spectra

(i) The Oxidised Native Fe(III) Protein

The absorption spectra of oxidised P-450 from pseudomonas putida are shown in Fig. 5 (*3*). While the bands at 416, 535 and 570 nm are typical of low-spin species the bands at 650 and especially the high absorption at 500 nm suggest the presence of a proportion of high-spin complexes (*52*). We presume that at room temperature the spectrum of P-450 is that of an equilibrium mixture of largely low-spin with some high-spin species. The evidence on the high-spin complex would be made more definitive by observations on the absorption spectrum at longer wave-

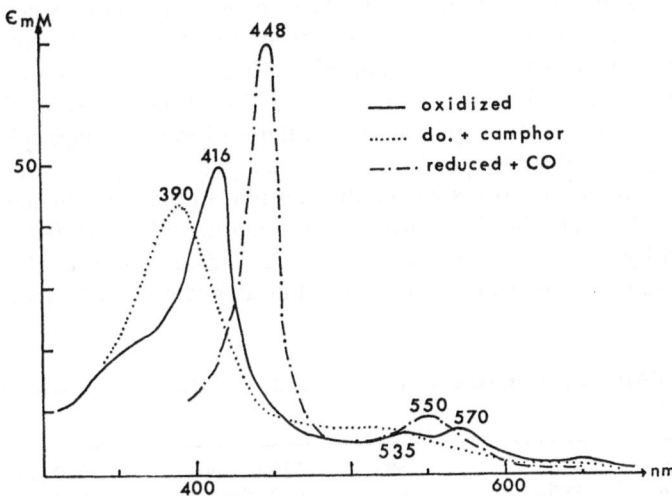

Fig. 5. The absorption spectra of P-450 and its complexes, after *Gunsalus* and co-workers

lengths, 700 to 1,500 nm, from which even the ligands of the high-spin form might be deduced (*52*).

The hemoglobin and myoglobin model thiol complexes have similar absorption spectra and they are largely low-spin from their magnetic moments. There are, however, differences between them and P-450: the Soret bands are at somewhat shorter wavelengths and the only long wavelength band appears at 700—800 nm in the model complexes. This band is not due to a high-spin component in the spectrum. A similar band is seen in the azide complexes of many largely low-spin Fe(III) hemoproteins and in cytochrome-*c* where there is considerable absorption at 690 nm even in the low-spin state. A possible explanation for all these observations is that a ligand or closely placed group perpendicular to the heme-plane is giving rise to a charge-transfer transition to the heme. Such groups must be reducing if the charge transfer band is to be at such low energy and it is therefore possible that some of the absorption of P-450 at around 650 nm arises from charge transfer transitions from an RS⁻ group in the close vicinity of the heme (*52*).

The above remarks apply strictly to the P-450 of bacteria. The P-450 of microsomes does not have quite the same spectrum and it depends upon the agent which has been used to induce it (*11, 25*). As we have seen when discussing the spin-resonance, the high-spin proportion in the low-spin/high-spin mixture varies from preparation to preparation. The characteristics of these two spin-states in heme-proteins generally are shown in Table 8, and a discussion of the spin-states of microsomal P-450 was given by one of us in 1967 (53). Since then various new preparations, *Estabrook et al.* (23), *Gaylor* and *Jefcoate* (24—26) have been made showing, by their absorption spectra, very different low spin: high spin ratios. Following the discussion of the previous paper (52) we can deduce that the induction process has modified a few amino-acids in the protein chain such that the heme-iron to protein-ligand bond-distances are changed by very small amounts e.g. 0.1 Å. This adaption of the protein to the inducing substrate looks rather like an antibody-antigen reaction.

Table 8. *Characteristic spectra of low-spin and high-spin hemeproteins, Fe(III),* $\lambda_{max}(nm)$

Low-Spin	415	535	570 (weak)	>1,000 (very weak)
High-Spin	390—410 (broad)	500	600—650	~1,000 (weak)

(ii) Binding of Substrates to the Fe(III) Protein

Binding substrates in the absence of oxygen produces two unusual spectroscopic effects. If the substrate is a non-coordinating molecule such as camphor, a sterol, or almost any other non-polar molecule, then the absorption rises at wavelengths below 400 nm and the Soret band becomes a rather broad band. The visible region of the spectrum shows a much increased high-spin content, i.e. absorption at 500 nm increases. On the other hand addition of substrates which could bind directly to the iron shift the Soret band in the opposite sense and the visible region shows less high-spin component. Roughly it would appear that the spin equilibrium is shifted towards high-spin by type I substrates and toward low-spin by type II substrates (53). The EPR signals of the low-spin forms are slightly shifted from those in the free protein, Table 4.

One difficulty associated with the understanding of the high-spin/low-spin equilibrium in these proteins is the lack of a temperature dependence. Many heme proteins have been examined over wide temperature ranges in order to discover the thermodynamic parameters associated with their spin-state equilibria — all show considerable changes in spectrum between room and liquid nitrogen temperatures. The absence of such a change for P-450 means either that ΔH for the change is zero, an unlikely situation, or that the system is not in thermal equilibrium. Apart from this problem there is that of the nature of the high-spin species with strong absorption below 400 nm. Before discussing model experiments disigned to reproduce the anomalies of the oxidised species of the protein we shall turn to the equally disturbing features of the absorption spectra of the reduced species.

(iii) The Reduced Protein

At first sight reduced native P-450 does not have an anomalous spectrum if it is compared with *model* heme complexes, Table 9. The Soret band is at 412 nm, there is a broad band in the visible at 550 nm, and the α and β-bands are not seen separately. This is like many a model high-spin Fe(II) complex. However if this spectrum is compared with that of other hemoproteins we see that it is unusual amongst the oxidases and oxygen carrier proteins, Table 9. These proteins have Soret bands at very long wavelengths 430—440 nm and although the α-band (590 nm) is almost lost relative to the β-band (555 nm) these bands are identifiable. The known structure of hemoglobin and myoglobin, 5-coordinate high-spin Fe(II), indicate that this type of spectrum is due to the peculiar state of the Fe(II) heme in these proteins. In no model has it proved possible to make 5-coordinate high-spin Fe(II) heme complexes with the possible exception of Wang's polymers.

Table 9. *Typical Fe(II) protoheme spectra (nm)*

Ligands	γ	β	α
$(H_2O)_2$	416	555	(585)
$(NH_3)_2$	415(88)	522(20)	553(34)
(pyridine)$_2$	420(190)	526(17.5)	557(34.2)
(cyanide)$_2$	430(100)	540(20)	570(11)
$H_2O.\,CO$	415(158)	543(15)	573(15)

The figures in parentheses are the millimolar extinction coefficients.

Now on addition of many ligands to reduced P-450 a remarkable change in the spectrum is seen such that the Soret-band intensifies ($\varepsilon \simeq 10^5$) and moves to very long wavelength, $\sim$450 nm, no α band appears, and the β-band is seen at around 550 nm, Table 2. This type of spectral change does not occur with all ligands. In some cases the Soret band moves to the normal low-spin position $\sim$420 nm and the α and β bands, at 540 and 570 nm respectively, are of equal intensity, Table 2. This is a more or less normal hemochrome spectrum. In yet other cases both these two forms are seen in equilibrium and the balance between them can be adjusted by detergents and pH changes. These examples have been described in an earlier section and correspond to CO, CN$^-$ and isonitrile complexes respectively. In the case of the ligands such as CO which shift the Soret band to 450 nm addition of denaturing reagents such as detergents and of certain mercurials change the spectrum to a normal low-spin form called P-420. The balance between 420 and 450 nm spectrum is *not* a property of a special ligand. It can arise when CO, RNC, or amines are bound to the heme. We shall see too that species with unusual Soret bands can be produced in many model systems so that it is not just a special protein-heme problem.

Closer inspection of the reduced P-450 spectrum in the light of these observations shows that there are additional anomalies relative to the supposedly simpler model systems. The extinction coefficient of the Soret band is only about 50—70 $\times$ 10^3, compare Table 9 where data for model complexes are presented, and there is very high absorption around 440—460 nm. This could arise from a very broad Soret absorption band or, more likely, it could be that there is more than one type of simple reduced P-450 and that one form of it absorbs at around 450 nm. The absorption in the visible region is also very broad indeed. We return to these features later.

At the present time we have no certain knowledge of the state of the heme in these 450 nm species. We do not know if there are heme aggregates although they are unlikely. It is therefore reasonable to look at systems where the haem is aggregated as well as those where it is not in order to see how the absorption spectra can be mimic-ed. It seems reasonable to assume that the iron is low-spin in the carbon monoxide, isocyanide, and nitric oxide complexes as no high-spin iron complexes of this type are known. In the high-spin or low-spin state it may be that the thiol is weakly bound, if at all, for Fe(II) heme in models or in hemoglobin does not bind to thiols. In an attempt to understand these spectra we shall use a semi-empirical approach based on the theoretical discussion in the previous article (*52*) and elaborated in what follows immediately. Only Fe(II) complexes will be analysed as the Fe(III) proteins have been previously examined (*52*).

IV. Theoretical Considerations of the Spectra

a) Energies of Transitions: Reduced Species

We shall consider firstly whether or not a low-spin system of a monomeric Fe(II) could give rise to the 450 nm spectrum.

The nature of the two electronic transitions of the porphyrin from the ground state orbitals, a_{1u} and a_{2u}, to the excited orbital, e_g, is such that the transitions move to longer wavelength the better the electron acceptor power of the substituents or the better the electron donor power of the nitrogens of the porphyrin, see reference (52). Because of configuration interaction (see reference (72), Fig. 7) this rule will only apply to the *baricentre* of the transitions and not necessarily to both the $\alpha\beta$ and the γ regions. If we consider the simplest metal derivatives, Na, K, Mg, Ca, Sr, Ba, Zn and Cd we see that the expected changes of wavelength occur (54), Fig. 6. The complexes with the greatest electron density on the

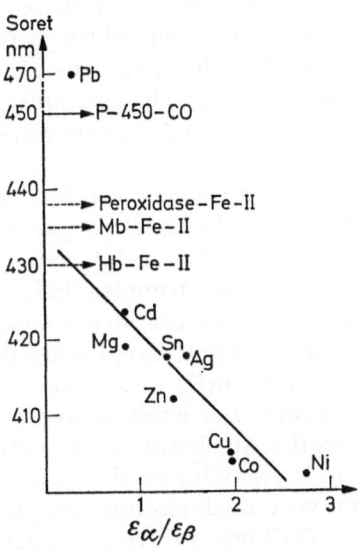

Fig. 6. The postions of the Soret band (γ) and the relative intensities of the visible absortion bands in the spectra of a numer of different metal porphyrins

nitrogens are $K > Na > Ba > Sr > Ca > Mg > Zn = Cd$. The discussion of mercury, tin, lead, silver(I) and all transition metal complexes has been excluded for reasons which will become apparent below. On binding zinc or cadmium porphyrin with an extra (fifth) ligand there is a marked band shift to longer wavelength, Table 10. Nuclear magnetic resonance and crystal structure studies have shown that the complexes of Zn(II) porphyrins are five coordinate with the metal lying well out of the plane of the porphyrin. The stronger the coordination of the fifth ligand to the metal the more the Soret band moves to longer wavelengths and it seems highly likely that the metal moves further out of plane and leaves a

Table 10. *Soret band changes on changes to 5-coordinate (55)*

	Zn	Cd	Fe	Co
4-coordinate i.e. in CCl$_4$	411.5	414.0	416	403
+ Dioxane 5-coordinate	417.5	422.5	?	414.5
(+ py)	422.5	432.5	430—40 (proteins)	

higher electron density on the nitrogens of the porphyrin. If we consider next a really large cation, even one which is of quite high electron affinity such as Pb or Hg, then the metal must sit on top of the porphyrin, i.e. a long way out of plane. The consequences in the absorption spectrum are seen in a very large shift to long wavelength, Fig. 6. This is also seen in the case of Ag(I).

Amongst transition metals we must consider the additional effect of the '*d*' orbitals. Referring to the previous paper mixing with charge-transfer transitions to the high-valent metal, high-spin Fe(III) or Mn(III), can be very strong so that the spectrum is totally distorted. Thus strong electron-acceptor metals must be considered separately. If we restrict consideration to the weaker electron acceptor metals in the series Mn(II), Fe(II), Zn(II), Co(II), Ni(II), Cu(II) we see that (*54*), much as expected, increasing σ-acceptor power, i.e. lower charge on the nitrogen forces the baricentre of absorption to shorter wavelengths, Fig. 6. The effect of 5-coordination (as in myoglobin) on the high-spin Fe(II) is to make this cation appear as a very weak electron acceptor i.e. the Soret band shift from symmetrical (415 nm) to five coordination (435 nm) is very similar to that for Zn(II) and Cd(II). The shift can be as large as to 440 nm in cytochrome *c* peroxidase and these complexes are therefore comparable with the 450 nm spectra of P-450 complexes.

Before going further it is as well to state the postulate that non-oxidising metal cations will shift the Soret band to longer wavelengths the more the porphyrin-nitrogen to metal bonds become ionic. When a metal sits out of the plane of the porphyrin the bond-length metal to nitrogen must be lengthened and it is reasonable to assume that there is a greater ionic character in the binding. Thus out-of-plane binding *for these cations* moves the Soret band to long wavelengths. We shall now provide additional experimental evidence in support of this postulate.

Our conclusions on the effect of 5-coordination upon the position of the Soret band of *reduced* metal species can be confirmed by reference to *Yonetani* and *Asakura's* work (*56*) on the spectra of Mn(II) hemes after incorporation of manganese porphyrins into different proteins. In these proteins 5-coordination of the Fe(II) protoporphyrin is known (myoglobin) or suspected (cytochrome-*c* peroxidase and horse radish peroxidase). Table 11 reproduces some of the data. The unbound Mn(II) porphyrins have Soret absorption bands in the region 415—420 nm while the bound hemes have bands at 426—440 nm. The largest shifts, with protoporphyrin, are as much as 20 nm, but no comparable change is seen in the Mn(III) spectra. (The switch from the parent model complex to the protein also causes a sharp drop in the α/β intensity ratio, see below).

Table 11. *Light absorption maxima of manganese porphyrin (56)*

complexes	γ	β	α
Mn(II)protoporphyrin(A)	420	547	577
mesoporphyrin(B)	416	543	573
A in Cytochrome-c peroxidase	439	563	595
B in Cytochrome-c peroxidase	428(90)	552(12)	583(6)
A in myoglobin	438	560	595
B in myoglobin	427(125)	552(16)	584(5)

Millimolar extinction coefficients in parentheses.

We next turn to an extensive study of the Ni(II) complexes for unlike Mn(II) and Fe(II) protoporphyrins the simple complexes are low-spin. *McLees* and *Caughey (57)* have shown that the addition of bases to nickel porphyrins causes a Soret band shift of 30 nm (410 → 440) with a concommitant change to high-spin, Table 12. The authors' discussion of this shift follows earlier views *(58)* in suggesting that the band shift is due to a fall in the electron acceptor power of the metal, greater ionic character, when it changes from low to high-spin. (At the same time as the Soret band shifts to longer wavelength the α/β band ratio decreases much as was observed above in the extended series of metal complexes and as is observed for P-450, (see below)).

Table 12. *Spectroscopic data for Ni(II) 2,4-substituted deuteroporphyrin IX dimethyl esters (57)*

	Low-Spin		High-Spin (pyridine)	
2,4 Substituent	Soret (λ)	α/β	Soret (λ)	α/β
Hydrogen	391	2.5	418	0.4
Acetyl	417	2.5	448	0.4
Vinyl	401	—	431	—

It is important to consider next the shift from high-spin to low-spin model Fe(II) hemes for the low-spin is both a better σ-electron acceptor and π-electron donor, Table 9. The baricentre is moved to shorter wavelengths with ligands such as ammonia and pyridine. However in the case of CO, RNC, NO the change is opposite and quite contrary to expectation for these ligands should have further *reduced* the electron density on the nitrogen. This leads us to propose that a very strong ligand on one side of the iron acts rather like 5-coordination. Perhaps the most important point to stress is that starting from simple model complexes the effect of strong ligands such as cyanide and carbonmonoxide is to move the absorption spectrum to longer wavelengths. The anomalous behaviour of hemoglobin, myoglobin, peroxidase, cytochrome-*c* peroxidase, tryptophan pyrrolase, where the movement is in the opposite direction is due to the five coordination of the high-spin Fe(II) and not due to any peculiarity of the carbon monoxide complexes.

b) Intensity of Spectra of Reduced Species

The intensity of the α absorption band depends upon the position in the configurational interaction diagram, rather than the baricentre of absorption. We can again consider the donor/acceptor character of the metal and its ligands separately.

Substituents in heme which lead to better electron acceptor properties, e. g. aldehyde groups as in heme *a*, give a grossly intensified α band. Substituents on the meso-carbon bridges alter the relative energy of the a_{1u} and a_{2u} orbitals so that the configurational interaction is rather different in say tetraphenyl porphyrin from that in porphyrin. It is known that aza substitution at these bridges as in phthalocyanines virtually removes the configurational interaction so that the αβ region absorbs equally with the γ. Metals affect the α/β intensity ratio in quite different ways depending upon whether the porphyrin is simple or is a tetraphenylporphyrin. In porphyrin itself, and taking A-subgroup and B-subgroup metals before transition metals, the α/β ratio increases with charge on the nitrogen indicating that the compounds are situated to the left of the centre of the configurational interaction diagram. Intensity changes are in the opposite sense for tetraphenylporphyrins (59) suggesting that the complexes are to the right of the centre of the diagram, Fig. 7. reference (72). Thus good acceptor metals enhance the α-band of porphyrins but diminish the α-band of tetraphenyl porphyrins. Transition metals additionally can couple charge transfer transitions, metal to ligand, with the porphyrin transitions. This will further enhance the intensity of the α-bands as seen in Co(II) Ni(II) Pd(II) and low-spin Fe(II) complexes. Ligands binding to metals can enhance (donors) or reduce (acceptors) the α-band intensity (58). For example increase of the π-bonding character of the axial ligand is found to lower the intensity of the α-band: the greater is the electron acceptor character of the new ligand the lower the α/β ratio:

$$CO, O_2 < CN^- < NH_3, \text{ pyridine.}$$

The absence of a ligand, as in the 5-coordinate myoglobin removes virtually all intensity from the α-band.

A similar effect can be generated in low-spin Fe(II) protoheme complexes by the addition of strong donor ligands. Such ligands will increase the electron density on the metal, much as has been shown in the corrin series, and should therefore decrease the strength of the metal-porphyrin bonds. Thus CN^- always reduces the α/β band ratio. The strength of the effect has been shown by Keilin in the dicyanide when the Soret band is at 438 nm and the α/β ratio is as low as 0.6. This effect is very like the cases described above of the Mn(II) porphyrin proteins (5-coordinate) and the change from low to high-spin in the nickel(II) porphyrin complexes.

Putting all the data on intensities together with that on the position of the Soret band we expect a relationship where the α/β intensity ratio decreases as the wavelength of the γ-band increases. Figure 6 shows how well this relationship is obeyed. It immediately suggests that the carbon monoxide complex of P-450 is an iron-heme protein in which iron is bound to no other ligand or a very weak ligand and is well out of the heme plane. Not only is the α-band virtually absent but the Soret band in many P-450 complexes is very broad, suggesting that it easily moves from one conformation to another. This is a very different situation from the usual bands of cytochromes c and b.

V. Model Complexes and Aggregation

a) Aggregates of Reduced Species

The greatest difficulty in the study of model, nonprotein, complexes is the correct identification of species. The monomolecular complexes which have been examined in organic solvents have spectra such as that shown in the previous article (52). As stressed throughout that article there would appear to be a valid and satisfactory theory of these spectra. However, there are circumstances in which quite different spectra are seen.

143

In the presence of critically controlled concentrations of isonitriles and pyridines various heme spectra have been obtained with Soret maxima close to 450 nm (60). Moreover the effects of detergents and pH on these spectra are somewhat similar to those seen with P-450 — there is a shift in the absorbtion towards 420 nm. It is clear that these unusual spectra are due to polymerisation and this has led many authors to suppose that the heme in P-450 is polymerised. (In the polymerised systems a carbon-monoxide pyridine complex with an absorption at 450 nm can even be produced.)

The following points strongly indicate that aggregation is not present in P-450

a) the chemical analysis indicates one heme per 50,000 i. e. one heme per protein
b) the EPR signals do not indicate interaction
c) the P-420 system is normal and it is difficult to visualise association and dissociation in a protein pocket.

While we cannot reject the possibility that polymers are present we can offer an alternative explanation for the parallel between the spectra of model polymers and monomeric P-450. The out-of plane character of the iron, which we show could well be responsible for the properties of P-450, could very well be induced by polymerisation in the presence of low-ligand concentrations.

b) Aggregates of Fe(III)-Hemes

The theory of the Fe(III) heme spectra has been given in the previous article (52) and in particular the difference between the absorption spectra of high-spin and low-spin species has been stressed. The application of this theory to some proteins has also been described in that article but its purpose was mainly to draw attention to normal spectra. Here we shall point to a number of anomalous spectra especially concerning the movement of the Soret band to much shorter wavelengths than 400 nm. There is a simultaneous notable broadening of this band and a fall in its extinction coefficient. Such effects have frequently been seen in simple model systems and so we deal these first.

Clezy and *Morrell* (61) discuss the appearance of different absorption spectra for a given porphyrin in different solvents and the effect of changing the substituents at the porphyrin perifery. In acetic acid there was only a single Soret maximum at 410—425 nm while in acetone, chloroform, ether and benzene a double Soret band was seen in the region 350—390 nm. It is probable that only one of these bands is really a Soret band and that the higher energy band is probably a forbidden transition. In a similar study, *Whitten, Baker,* and *Corwin* (62) examined the absorption spectrum of Fe(III) mesoporphyrin (IX) dimethylester in different solvents. They and *Brown* and *Slantzke* (*63*) report that in non-polar non-coordinating solvents the Soret band moves below 400 nm. Some of our own measurements on similar systems are included in Table 13 and show too that solvents such as CH_2Cl_2 differ markedly from say alcohols in their effect upon heme —Fe(III) absorption spectra. It is usual to associate the broad, short wavelength absorption bands with aggregated systems, and coordinating solvents are presumed to cause the aggregates to dissociate to monomers.

Table 13. *Soret bands of protoporphyrin IX dimethylester iron(III) chloride in different solvents*

Solvent	λ_{max}(nm)	
Methylene dichloride	385	non-coordinating solents
Ethylacetate	378	
Ethylchloracetic ester	382	
Bornylacetate	385	
Methanol	400	weak coordinating solvents
Cyclohexanol	400	
Dimethylformamide	410	

An alternative possibility discussed by *Brown* and *Slanzky* (*63*) and by ourselves (*52*) is that the iron(III) porphyrin becomes 5-coordinate, the iron being out of plane, and the porphyrin becomes ruffled. The evidence for this is based partly on solution studies and partly on X-ray crystal structure studies. *König* (*64*) and *Hoard* (*65*) and their coworkers have shown that iron porphyrin chlorides are five-coordinate, square-pyramids in which the iron lies some 0.5 Å out of the plane of the ring. Even in so-called six-coordinate systems such as the H_2O—Fe(III)—OH^- complex of tetraphenylporphyrin the iron is about 0.4 Å from the plane. In' Fe(III) mesoporphyrin methoxide the iron is also out of the plane by about 0.5 Å. In these crystal structures there is no suggestion of special bonding between the heme units so that if they aggregate it seems likely that no new bonds are formed to the iron atoms — they are likely to remain strictly 5-coordinate. On the other hand placing the compounds in coordinating solvents will lead to dissociation of the gegen-anion, chloride, and can really give six-coordinate monomers. It can well be then that the lowering of the Soret wavelength is not due to the aggregation but is seen in aggregates because here the iron(III) remains 5-coordinate and far from the heme plane. We need now to show that similar low-intensity high-energy Soret bands can arise in heme polymers and proteins where the structure prevents heme-heme interaction.

H. A. O. Hill, A. Röder, and R. J. P. Williams

Table 14 gives data on the effect of acidification on the Soret band of a number of proteins. Accompanying the drop in the wavelength there is a decrease in intensity of the band. Directly comparable with this observation is the study of hemes incorporated into polymers by *Weightman (66)*. His results show that the Soret band moves well below 400 nm

Table 14. *Soret-band positions (nm) in hemoproteins at different pH*

	$pH = 7.0$	$pH = 0.6$
Peroxidase	407	397
Catalase	405	397
Hemoglobin	405	398
Myoglobin	408	397

except in the presence of added base. A much more striking example of these effects is provided by the work of *Horio* and *Kamen (67)* who studied the spectra of *rhodospirillum rubrum* cytochrome cc¹. The spectra and their dependence on pH are shown in Fig. 7. It will be seen that at low pH the spectra show a Soret band well below 400 nm, the iron(III) is high-spin and that as the pH is raised this changes to a low-spin spectrum with a Soret band maximum above 400 nm.

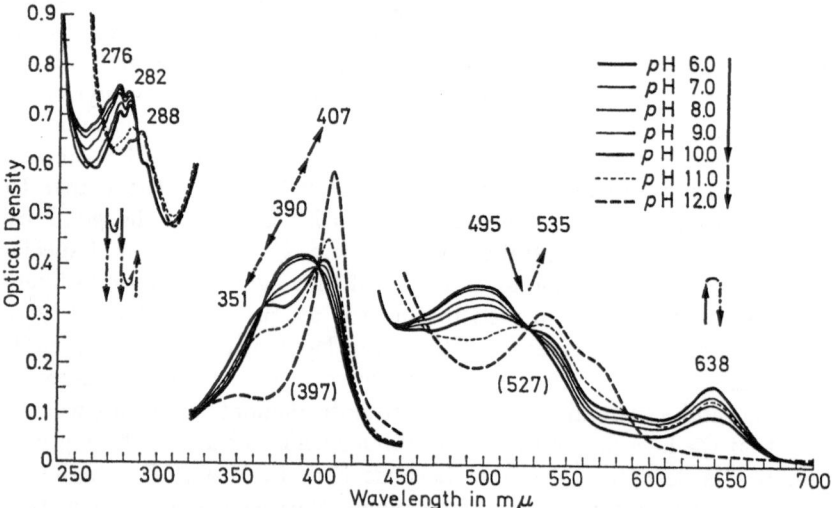

Fig. 7. The absorption spectra of oxidised rhodospirillum rubrum cytochrome cc' (oxidised) with change of pH

146

VI. Summary of Spectra

It appears from the discussion given above that we must postulate that distortion of the heme and concommittant 5-coordination of the iron lead to quite different effects in the porphyrin absorption bands in the two oxidation states. Theoretical justification for such a statement is not presented but we note that whereas the lowest orbitals of iron(III) are strong acceptors, the lowest orbitals of iron(II) are strong donors, no matter which spin states are being considered. In the absence of theoretical justification we note the following further observations.

1. Polymerisation of iron(II) and iron(III) porphyrins leads to opposite shifts in the Soret bands, Fe(III) to shorter and Fe(II) to longer wavelengths.

2. In such enzymes as peroxidase the Fe(II) Soret band is at very long wavelengths (435 nm) while that of Fe(III) is at rather short wavelengths (400 nm).

3. In that protein, rhodospirillum rubrum cytochrome cc[1], which shows the most remarkable shift of the Soret band to short wavelengths for the Fe(III) state at low pH there is an opposite shift to long wavelengths in the Fe(II) state, Fig. 8 (67). In fact this cytochrome behaves remarkably like P-450 in its isonitrile complexes.

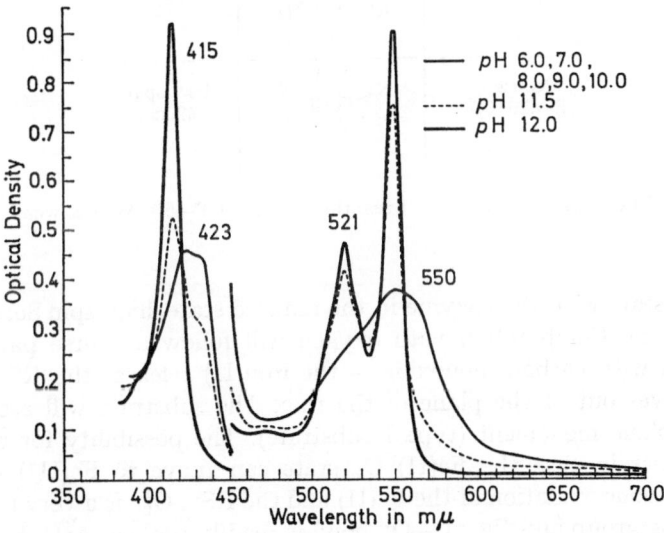

Fig. 8. The absorption spectra of reduced rhodospirillum rubrum cytochrome cc′ with change of pH. This figure and figure 7 are reproduced with permission (67)

VII. General Conclusions and Reaction Mechanisms

From all the observations on the physical properties of P-450 we conclude that a scheme of the states of this protein can be drawn up as in Fig. 9, which is self-explanatory. In the light of this figure we can go on to discuss the mode of action of P-450.

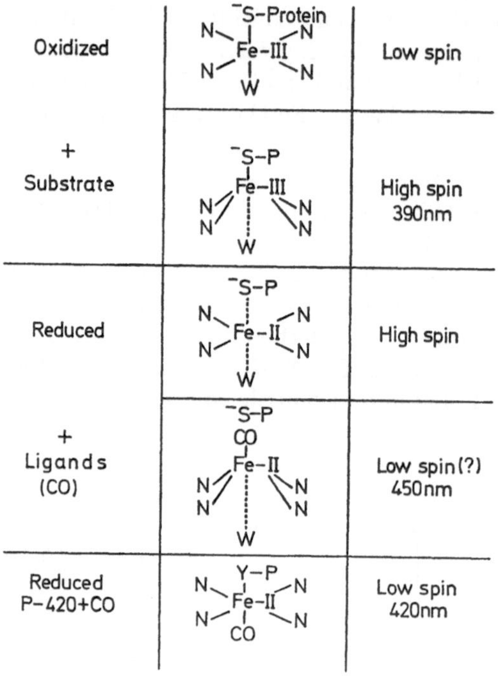

Oxidized	$^-$S–Protein N⟍|⟋N Fe–III N⟋|⟍N W	Low spin
+ Substrate	$^-$S–P | Fe–III N⟋⋮⟍N N⟋⋮⟍N W	High spin 390nm
Reduced	$^-$S–P N⟍⋮⟋N Fe–II N⟋⋮⟍N W	High spin
+ Ligands (CO)	$^-$S–P CO | Fe–II N⟋⋮⟍N N⟋⋮⟍N W	Low spin(?) 450nm
Reduced P–420+CO	Y–P N⟍|⟋N Fe–II N⟋|⟍N CO	Low spin 420nm

Fig. 9. The suggested scheme of reaction states of P-450. W is a weak donor

We start with the enzyme in the reduced state, high-spin Soret band at 415 nm. Combination with oxygen will follow a course parallel to reaction with carbon monoxide — the iron(II) releases the RS$^-$ group and moves out of the plane of the ring. The substrate will assist this out of plane movement (type I substrate). The possibility for reaction now arises in that the Fe(II) O$_2$ state can move to Fe(III) O$_2^{2-}$ by simultaneous oxidation of the Fe(II) and the RS$^-$. O$_2^{2-}$ is a very powerful attacking group and the $^-$O—O$^-$ bond is readily broken. Attack on substrate, S, giving SOH is possible and the iron centre is left as RS. FeO which, by transfer of electron from putaredoxin at this stage or earlier

148

will revert to RS⁻ FeO. Subsequently electron transfer steps from NADH through the putaredoxin will reduce successively to RS⁻ Fe(III) and finally to RS⁻ Fe(II) and H_2O. Thus the mixed function action is established by the electron transfer chain, Fig. 1.

If our postulates are correct the most interesting feature of P-450 is the manner in which the protein has adjusted the coordination geometry of the iron and then provided near-neighbour reactive groups to take advantage of the activation generated by the curious coordination. *Vallee* and *Williams* (68) have observed this situation in zinc, copper and iron enzymes and referred to it as an entatic state of the protein. It is also apparent that some such adjustment of the coordination of cobalt occurs in the vitamin B_{12} dependent enzymes. As a final example we have looked at the absorption spectra of chlorophyll for its spectrum is in many respects very like that of a metal-porphyrin. This last note is intended to stress the features of chlorophyll chemistry which parallel those of P-450.

VIII. The Nature of Activated Chlorophyll

In the chloroplast and in the active centres of chromatophores of bacteria there are special sites for the magnesium chlorophyll and bacteriachlorophyll respectively, which absorb light at very different wavelengths from the normal chlorophyll. These sites are often referred to under numberings (like that given to P-450) P-750 and P-870. These active sites could well be molecules of chlorins in which the metal has been forced to move far out-of-plane in order to accommodate steric hindrance between other parts of the chlorin molecule and the binding protein side-chains and in order to make a reasonable fifth bond to a protein side chain. The NMR data of *Katz* and coworkers (69) and more recently of *Ballschmiter* and *Katz* (70) and the crystallographic data of *Tulinsky* (71) all point to an out-of-plane structure for Mg(II) complexes. Thus the specially activated P-450 iron complexes may well be only one example of a general activation of porphyrins and chlorins comparable with metal activation in metallo-enzymes.

One of us (A. R.) wishes to thank the European Molecular Biology Organisation for a grant which enabled him to take part in this work. This work was supported by the Medical Research Council, England.

IX. References

1. *Omura, T., Sato, R.:* J. Biol. Chem. *239*, 2370 (1964). — *Omura, T., Sato, R., Cooper, D. Y., Rosenthal, O., Estabrook, R. W.:* Federation Proc. *24*, 1181 (1965).
2. *Klingenberg, M.:* Arch. Biochem. Biophys. *75*, 376 (1958). — *Garfinkel, D.:* Arch. Biochem. Biophys. *77*, 493 (1958).
3. *Hedegaard, F., Gunzalus, J. C.:* J. Biol. Chem. *240*, 4038 (1965). — *Appleby, C. A.:* In: Symposium on Cytochromes, p. 357; ed. M. Kamen and K. Okunuki. Tokyo: University Press 1967.
4. *Mason, H. S.:* Ann. Rev. Biochem. *34*, 595 (1965).
5. *Mashimoto, Y., Yamano, T., Mason, H. S.:* J. Biol. Chem. *237*, 3843 (1962).
6. *Katagiri, M., Ganguli, B. N., Gunzalus, J. C.:* Federation Proc. *27*, 525 (1968).
7. *Estabrook, R. W., Cooper, D. Y., Rosenthal, O.:* Biochem. Z. *338*, 741 (1963).
8. *Omura, T., Sato, R., Cooper, D. Y., Rosenthal, O., Estabrook, R. W.:* Federation Proc. *24*, 1181 (1965).
9. *Hedegaard, J., Gunsalus, J. C.:* J. Biol. Chem. *240*, 4038 (1965).
10. *Ernster, L., Orrenius, S.:* Federation Proc. *24*, 1190 (1965).
11. *Remmer, H., Schenkman, J. B., Estabrook, R. W., Sasame, H., Gillet, J., Narasimkulu, S., Cooper, D. J., Rosenthal, O.:* Mol. Pharmacol. *2*, 187 (1966).
12. *Mason, H. S., North, J. C., Vanneste, M.:* Federation Proc. *24*, 1172 (1965).
13. *Ichikawa, Y., Yamano, T.:* Arch. Biochem. Biophys. *121*, 742 (1967); Biochim Biophys. Acta *131*, 490 (1967); *147*, 518 (1967).
14. *Imai, J., Sato, R.:* J. Biochem. (Tokyo) *62*, 239 (1967); *63*, 270 (1968).
15. *Fujita, T., Itagaki, E., Sato, R.:* J. Biochem. (Tokyo) *53*, 282 (1963).
16. *Hansch, C., Hicks, K., Lawrence, G. C.:* J. Am. Chem. Soc. *87*, 5770 (1965).
17. *Omura, T., Sato, R.:* J. Biol. Chem. *239*, 2370 (1964).
18. *Cooper, D. J., Levin, S., Narasimbubu, S., Rosenthal, O., Estabrook, R. W.:* Science *147*, 400 (1965).
19. *Nisihibayashi, H., Sato, R.:* J. Biochem. (Tokyo) *61*, 491 (1967).
20. — — *Omura, T.:* Biochim. Biophys. Acta *118*, 651 (1966).
21. *Imai, Y., Sato, R.:* J. Biochem. (Tokyo) *64*, 147 (1968); *62*, 464 (1967).
22. *Kinoshita, T., Horie, S.:* J. Biochem. (Tokyo) *61*, 26 (1961).
23. *Remmer, H., Estabrook, R. W., Schenkman, J.:* Arch. Pharmakol. Expt. Pathol. *259*, 98 (1968).
24. *Jefcoate, C. R. E., Gaylor, J. C.:* J. Am. Chem. Soc. *91*, 464 (1968).
25. — — *Calabrese, R. L.:* Biochemistry *8*, 3455 (1969).
26. — — Biochemistry *8*, 3464 (1969).
27. *Omura, T., Sato, R., Cooper, D. Y., Rosenthal, O., Estabrook, R. W.:* Federation Proc. *24*, 1181 (1965).
28. *Mitani, F., Horie, S.:* J. Biochem. (Tokyo) *65*, 269 (1969).
29. *Hashioto, Y., Yamo, T., Mason, H. S.:* J. Biol. Chem. *237*, 3843 (1962),
30. *Mitami, J., Horie, S.:* J. Biochem. (Tokyo) *66*, 139 (1969).
31. *Ichikawa, J., Yamano, T.:* Biochem. Biophys. Acta *153*, 753 (1968).
32. *Schenkman, J. B., Greim, H., Zange, M., Remmer, H.:* Biochem. Biophys. Acta *171*, 23 (1969).
33. *Hildebrandt, A., Remmer, H., Estabrook, R. W.:* Biochem. Biophys. Res. Commun. *30*, 607 (1968).
34. *Ehrenberg, A.:* Arkiv Kemi *19*, 119 (1962).
35. *Gibson, J. F., Ingram, D. J. E., Schonland, D.:* Discussions Faraday Soc. *26*, 72 (1958).
36. *Torii, K., Ogura, Y.:* J. Biochem. (Tokyo) *64*, 171, 1968; *65*, 825 (1969).
37. *Murakami, K., Mason, H. S.:* J. Biol. Chem. *242*, 1102 (1967).

38. *Mason, H. S., North, J. C., Vanneste, M.:* Federation Proc. *24*, 1164 (1965).
39. *Röder, A., Bayer, E.:* Angew. Chem. *79*, 274 (1967).
40. *Jefcoate, C. R. E., Gaylor, F. C.:* Biochemistry *8*, 3464 (1969).
41. *Mori, K. Y., Mason, H. S.:* J. Biol. Chem. *240*, 2659 (1965).
42. *Yonetani, T., Schleyer, H., Ehrenberg, A.:* In: Magnetic Resonance in Biology, p. 151; eds. A. Ehrenberg, B. G. Malmstrom, and T. Vanngard. New York: Pergamon Press 1967.
43. *Bois-Peltroratsky, R., Ehrenberg, A.:* European J. Biochem. *2*, 361 (1967).
44. *Bayer, A., Röder, A.:* Europ. J. Biochem. *11*, 89 (1969).
45. *Bayer, E., Hill, H. A. O., Röder, A., Williams, R. J. P.:* Chem. Commun. 109 (1969).
46. *Hill, H. A. O., Röder, A., Williams, R. J. P.:* Naturwissenschaften *57*, 69 (1970).
47. *Röder, A.:* Doctorate thesis, Tübingen.
48. *Hollocher, T. C., Buckley, L. M.:* J. Biol. Chem. *241*, 2976 (1966).
49. *Banerjee, R., Alpert, Y., Leterrier, F., Williams, R. J. P.:* Biochemistry *8*, 2862 (1969).
50. *Rumen, N. M.:* personal communication.
51. *Hill, H. A. O., Macfarlane, A. J., Mann, B., Williams, R. J. P.:* Chem. Commun. 123 (1968).
52. *Smith, D. W., Williams, R. J. P.:* Struct. Bonding *7*, 1 (1970).
53. *Williams, R. J. P.:* In: Conference on Cytochromes, p. 393; ed. M. Kamen and K. Okunuki. Tokyo: University Press 1967.
54. *Caughey, W. S., Deal, R. M., Weiss, C., Gouterman, M.:* J. Mol. Spectry. *16* 451 (1965).
55. *Phillips, J. N.:* In: Comprehensive Biochemistry, Vol. 9, p. 34; eds. M. Florkin and E. H. Stotz. Amsterdam: Elsevier 1963.
56. *Yonetani, T., Asakura, T.:* J. Biol. Chem. *244*, 4580 (1968).
57. *McLees, B. D., Caughey, W. S.:* Biochemistry *7*, 642 (1968).
58. *Williams, R. J. P.:* Chem. Rev. *56*, 299 (1956).
59. *Corwin, A. H., Erdman, J. G.:* J. Am. Chem. Soc. *68*, 473 (1946). — *Barnes, J. W., Dorough, G. D.:* J. Am. Chem. Soc. *72*, 4045 (1950).
60. *Holden, H. F., Lemberg, R.:* Australian J. Exptl. Biol. Med. Sci. *17*, 133 (1939). — *Keilin, J.:* Biochem. J. *45*, 440 (1949). — *Imai, Y., Sato, R.:* J. Biochem. (Tokyo) *64*, 147 (1967).
61. *Clezy, P. S., Morell, M.:* Biochem. Biophys. Acta *71*, 165 (1963).
62. *Whitten, D. G., Baker, E. W., Corwin, A. H.:* J. Org. Chem. *28*, 2363 (1963).
63. *Brown, S. B., Lantsky, I. R.:* Biochem. J. *115*, 279 (1969).
64. *König, D. F.:* Acta Cryst. *18*, 663 (1965).
65. *Hoard, J. C.:* In: Hemes and Hemoproteins, p. 9; eds. B. Chance, R. Estabrook, and T. Yonetani. New York: Academic Press 1966.
66. *Weightman, J.:* personal communication.
67. *Horio, T., Kamen, M. D.:* Biochim. Biophys. Acta *48*, 266 (1961).
68. *Vallee, B. L., Williams, R. J. P.:* Proc. Natl. Acad. Sci. *59*, 498 (1968).
69. *Katz, J. J., Strain, H. H., Leussing, D. L., Dougherty, R. C.:* J. Am. Chem. Soc. *90*, 784 (1968).
70. *Ballschmiter, K., Katz, J. J.:* Angew. Chem. *80*, 283 (1968).
71. *Timkovich, R., Tulinsky, A.:* J. Am. Chem. Soc. *91*, 4430 (1969).
72. *Braterman, P. S., Davies, R. C., Williams, R. J. P.:* Advan. Chem. Phys. *VII*, 359 (1964).

Received February 4, 1970

Cobalt(II) in Metalloenzymes.
A Reporter of Structure-Function Relations

S. Lindskog

Institutionen för Biokemi, Göteborgs Universitet and Chalmers Tekniska Högskola,
Göteborg, Sweden

Table of Contents

I. Introduction

In contrast to iron and copper, which dominate the scene of transition metal biochemistry and are components of a variety of metalloproteins, cobalt occupies a relatively modest niche in biology. A biological function of cobalt can only be said to have been established at the molecular level in a few cases involving *corrinoid coenzymes* (1).

In these compounds the cobalt atom is enclosed in a highly conjugated cobalamin structure and linked to an alkyl group via a metal-carbon bond. The B_{12} coenzymes are diamagnetic and can be regarded as complexes of cobalt(III) with a carbanion as a ligand (2). As this review will be limited to cases of direct metal-protein interactions the corrinoids will not be discussed further.

Most of the literature on cobalt(II) in biochemistry concerns its effects in various *metal-activated enzyme systems* (3). Many of the typical Mg^{2+}-activated enzymes can work with Co^{2+}, though usually at a low rate. In some other systems, where Mn^{2+} commonly is the best coenzyme, Co^{2+} gives high activities while other cations are less effective or inhibitory. A few enzymes, notably some metal-activated peptidases, are most efficient with Co^{2+}, but other metal ions are also functional. It is not believed, however, that Co^{2+} is an important enzyme activator *in vivo* (4).

As discussed in detail by *Malmström* and *Rosenberg* (5) and by *Williams* (6), the specificity of metal ion activation can not be directly correlated with any given property of the metal ions or simple model complexes, but must be a consequence of special structural features of the enzymic active sites. The fact that activation can be achieved with a set of different cations embodies the possibility of using the spectroscopic and magnetic properties of cobalt and other transition metal ions to obtain some of the desired structural information. These properties are generally very sensitive to changes in the coordination sphere and should therefore be valuable as probes of events taking place in the active sites on the binding of substrates and inhibitors or during the catalytic action (7).

One of the difficulties in applying this approach to metal-activated enzymes is inherent in their definition. The *stability* of the catalytically interesting complex is small and a non-negligible concentration of free ions must be present to achieve saturation. A mixture of complexed species will result, especially as there often is a tendency towards *non-specific binding to secondary sites*. However, the paramagnetism of Mn^{2+}, particularly the effects on the proton relaxation of coordinated water molecules, has been fruitfully exploited to give insight into the functions of several such enzymes (8). This problem is generally not encountered in the study of metalloproteins, as they are normally isolated as stable, well-defined complexes. Instead, there may be considerable difficulties in finding conditions where metal substitution is feasible, and the studies must often be restricted to the constitutive metal ion. However, this is not a serious limitation when the intrinsic metal is copper, iron or molybdenum, which are excellent built-in probes.

An important group of metalloenzymes contain Zn^{2+}, which has a complete set of $3d$ electrons and lacks the physical properties characterizing transition metal ions. In their pioneering work on *carboxypeptidase A, Vallee et al.* (9) showed that the zinc ion could be removed by dialysis against a chelating agent with a concomitant loss of activity, and that a significant restoration of activity was obtained with some other metal ions, including Co^{2+}. Substitution has later been achieved for several zinc enzymes. A general observation seems to be that cobalt generates catalytic activity, often approaching, and sometimes exceeding that of the native enzyme.

The purpose of the present review is to summarize how cobalt-linked absorption spectra and other physical properties have been utilized in attempts to elucidate relations between *structure and function in these enzymes*. The emphasis will be on *carbonic anhydrase* not only because it reflects the author's own interests, but mainly because it is the most extensively studied cobalt enzyme. Its environmentally-sensitive absorption spectrum has furnished essential information as to the role of the metal ion in the catalytic reaction. For other enzymes, the probe properties of cobalt are just beginning to be explored, but significant advances have recently been reported (7).

Some efforts have been made to interpret the spectroscopic and magnetic properties of cobalt enzymes in terms of *coordination geometry* and chemical identity of ligands. The basis of these attempts is a comparison with the corresponding properties of low-molecular weight complexes of known structure. A brief summary of relevant data on some "models" is given in the following section.

II. Structures, Absorption Spectra and Paramagnetism of Co(II) Complexes

A. Stereochemistry

The cobalt(II) ion preferentially forms octahedral complexes, but tetrahedral coordination is not uncommon. The tendency to form tetrahedral complexes is less marked than for Zn(II), but much stronger than for Ni(II) or Cu(II) (10). The actual geometry obtained with given ligands depends on both electrostatic and steric factors. For example, many amine and amino acid chelates of Co(II) are octahedral (*11, 18*) while halide anions form tetrahedral complexes (*12*).

Considerable deviations from exact tetrahedral symmetry are often observed, particularly in complexes with mixed and/or bulky ligands. The term *pseudotetrahedral* is sometimes used to indicate this kind of distortion in complexes having essentially the physical properties of tetrahedral coordination. Tetrahedral complexes tend to add solvent molecules in coordinating solvents such as pyridine or water, forming equilibrium mixtures with different coordination numbers. Ligand exchange is rapid in Co(II) chelates in contrast to the inert Co(III) complexes.

Many five-coordinated complexes of Co(II) are now known, having trigonal bipyramidal or square pyramidal structures. A much used type of ligand is a tetradentate tripod-like molecule with bulky groups attached to the bonding atoms, allowing coordination at only one additional position. In this case the symmetry is a more or less distorted trigonal bipyramid (*13*).

Square planar complexes of Co(II) are uncommon, but are formed with certain types of ligands (see the reviews by *Hare* (*14*) and by *Carlin* (*15*)).

The stereochemical preferences of Zn(II) and Co(II) are rather similar and, as pointed out by *Dennard* and *Williams* (*16*) and *Vallee* and *Williams* (*17*), the versatility of these ions in accepting different ligand geometries may be of importance for their function and interchangeability in metalloenzymes.

B. Absorption Spectra

The Co(II) ion has a d^7 configuration, and both high-spin and low-spin complexes exist. Studies of a large number of Co(II) complexes have shown that the low-energy (*d-d*) absorption spectra are so characteristic that the geometry of a complex can be reasonably well predicted from its spectrum. Detailed discussions of absorption spectra of Co(II) in different environments are given by *Carlin* (*15*) and by *Ciampolini* (*13*). Only a descriptive survey of typical spectral features will be given here. Charge transfer and ligand absorption bands will not be considered as these are less well understood and less readily generalized then *d-d* transitions.

Representative absorption spectra of Co(II) in *octahedral* and *tetrahedral fields* are shown in Fig. 1 and Fig. 2, respectively. The maximal molar absorbance of an octahedral complex is of the order $10 \ M^{-1} \ cm^{-1}$. A peak of even lower intensity is found in the near-infrared region. The bands usually have only little structure.

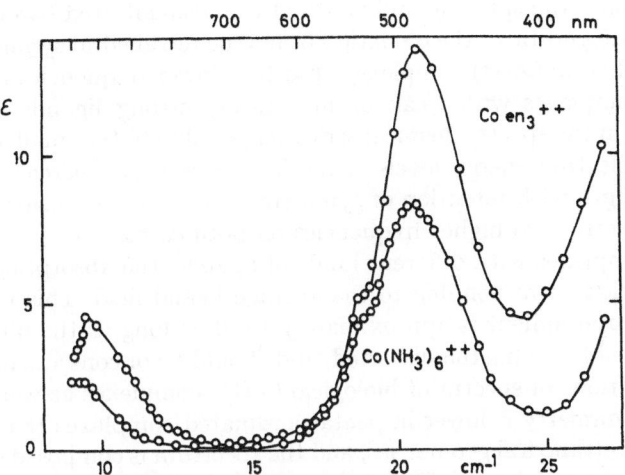

Fig. 1. Absorption spectra of $Co(NH_3)_6^{2+}$ and $Co(ethylenediamine)_6^{2+}$ in aqueous solutions containing 12 M NH_3 and 0.5 M ethylenediamine, respectively. From *Ballhausen* and *Jørgensen* (*18*)

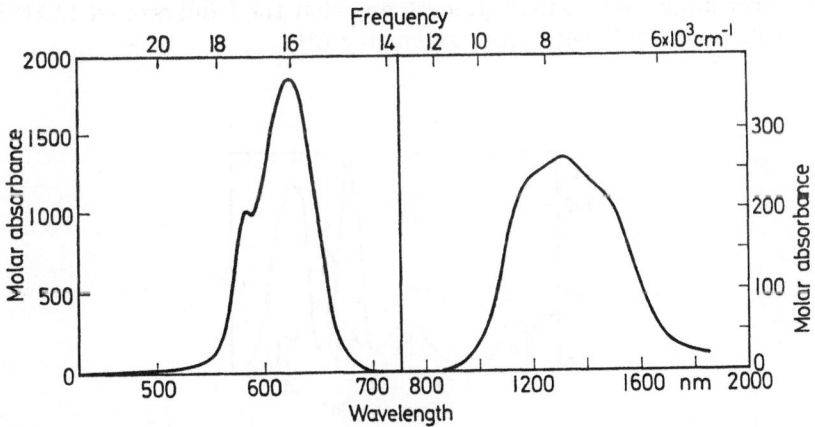

Fig. 2. Absorption spectrum of $Co(NCS)_4^{2-}$ in acetone solution. Note the difference in absorbance scale in this Figure and in Fig. 1. From *Cotton et al.* (*19*)

157

The (pseudo-)tetrahedral complexes are much more intensely colored. Maximal molar absorbancies range from several hundreds to over 1000 M^{-1} cm^{-1}. There is also a band in the near infrared with ε_{max} of the order 100 M^{-1} cm^{-1}. Both bands show a great deal of structure. Such splittings may be caused by spin-orbit coupling, but contributions from ligand fields of low symmetry are thought to be more important in this case. In fact, the splitting and width of the near-infrared band has been taken as diagnostic of the deviation from true tetrahedral symmetry (20).

Tetrahedral Co(II) complexes absorb at lower frequencies than octahedral complexes with weak or moderately strong ligands of similar positions in the spectrochemical series. This is due to the smaller splitting of the d-electron energy levels in the former case. An increase of ligand field strength with retention of symmetry and spin state leads to a shift of the spectrum to higher frequencies on both cases.

In complexes with different kinds of ligands, the absorption appears at a position corresponding to the average ligand field. This rule of the average environment is approximately valid as long as the mixed complex does not become too distorted, and should be of consequence for the interpretations of spectra of biological Co(II) complexes as well.

The symmetry is lower in pentacoordinated complexes than in octahedral or tetrahedral geometries, and the spectrum is composed of several widely separated bands. High-spin pentacoordinated Co(II) complexes have absorption bands with intensities intermediate between those of tetrahedral and octahedral species. A characteristic spectrum of a complex with trigonal bipyramidal symmetry is given in Fig. 3. Further examples can be found in a recent review by *Ciampolini* (13) and in references therein. *Ciampolini* points out that the band around 16 kK is rather insensitive to field strength and that the band around 12 kK is split in compounds with lower symmetry (21).

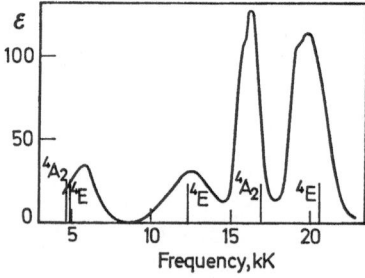

Fig. 3. Absorption spectrum of [CoBr(Me₆tren)]Br in CH_2Cl_2 solution. Me₆tren = $N[CH_2CH_2N(CH_3)_2]_3$. From *Ciampolini* (13)

Several pentacoordinated low-spin Co(II) complexes have been prepared. These compounds are often very intensely colored with molar absorbancies in the visible and near ultraviolet regions of a few times 10^3 M^{-1} cm^{-1} (13).

The low-energy d-d bands of planar Co(II) complexes have rather low intensities, but the visible spectra are often dominated by high-intensity charge-transfer and ligand bands (14).

C. Paramagnetism

Both high-spin ($S = 3/2$) and low-spin ($S = \frac{1}{2}$) Co(II) complexes are para-magnetic. In an octahedral symmetry, a very strong field is required before spin pairing occurs and only a few such complexes are known. On the other hand, practically all complexes of Co(III) are spin paired and diamagnetic (d^6). Some Co(II) chelates tend to combine with mole-cular oxygen to form diamagnetic compounds (see ref. 22). In a tetra-hedral field only the high-spin state is possible for the d^7 configuration (10, 15). Square planar Co(II) complexes are generally of the low-spin type, but one supposedly planar high-spin compound has recently been reported (23).

Due to spin-orbit coupling, the magnetic moments of Co(II) com-plexes are larger than the spin-only moments for one (1.73 Bohr magne-tons, B.M.) or three (3.87 B.M.) unpaired electrons, respectively. The influence of the ligand field on the contribution of the spin-orbit coupling to the magnetic moment is discussed in Carlin's review (15) and by Cotton et al. (12). In Table 1, the observed ranges for the magnetic mo-ments in different environments are shown.

Rather few *electron paramagnetic resonance (EPR) spectra* of Co(II) are reported in the literature. A summary of EPR parameters published through 1964 is given by Carlin (15). Due to short spin-lattice relaxation times, EPR signals of high-spin Co(II) are broad and hard to detect except at very low temperatures. A few biologically interesting compounds representing low-spin Co(II) in a tetragonal field have been studied. As an example can be mentioned vitamin B_{12r}, which is obtained on reduc-tion of aquocobalamin or by photolysis of 5'-deoxyadenosylcobalamin. Most of the eight hyperfine lines due to the nuclear spin 7/2 of ^{59}Co are resolved in the $g_{\parallel}$ region and, in addition, nitrogen superhyperfine struc-ture is observed and attributed to the interaction with the benzimid-azole residue (24). A similar spectrum was also obtained in a system containing the corrinoid coenzyme-dependent ribonucleotide reductase from *Lactobacillus leichmanni* (25).

Table 1. *Effective magnetic moments (μ_{eff}) for Co(II)-complexes of different geometries (12, 13, 15, 23)*

Structure	S	μ_{eff}, Bohr magnetons (B.M.) room temperature
Octahedral	3/2	4.7—5.2
	1/2	1.8—1.9
Tetrahedral	3/2	4.3—4.8[a]
Pentacoordinated	3/2	4.4—5.5[b]
	1/2	1.9—2.1
Planar	3/2	5.0[c]
	1/2	2.1—2.5

[a] $\mu_{eff} = 3.87 \, (1-4 \, \lambda'/\triangle)$, where λ' is the effective value of the spin-orbit coupling constant, averaging -148 cm^{-1} in the tetrahedral complexes (12).

[b] The high values (> 5.0 B.M.) represent a few complexes with relatively regular square pyramidal coordination. Distortions lead to lower values. Trigonal bipyramidal complexes have μ_{eff} in the range 4.4—4.8 B.M. See references given by *Ciampolini* (13) for details.

[c] One complex reported in ref. (23).

The paramagnetic properties of Co(II) have been utilized in some biochemical applications of nuclear magnetic resonance. Cobalt(II)-induced contact shifts were observed in lysozyme (26). A preferential binding of Co^{2+} to a single site presumably involving two carboxyl groups was deduced. This technique might become very informative in studies of metal ion-dependent enzyme systems.

III. Cobalt(II) in Metalloenzymes

A. Carbonic Anhydrase

Carbonic anhydrase was the first known example of a zinc-containing metalloenzyme (27). It is present in a large number of tissues in all vertebrates and in many invertebrates. It has also been found in the green tissues of plants and in some bacteria (28). The primary reaction catalyzed by the enzyme

$$CO_2 + H_2O \rightleftharpoons H^+ + HCO_3^-$$

is readily reversible at physiological pH and proceeds with an appreciable rate even in the absence of a catalyst. The enzyme is not directly involved in intermediary metabolism, but it plays a central role in the maintenance and rapid adjustment of the "acid-base balance". The most abundant sources are mammalian red blood cells, where carbonic anhydrase is present in concentrations much exceeding what is though to be necessary for its function in respiration.

Some other reactions, such as aldehyde hydration (29) and ester hydrolyses (30—33) are also catalyzed by the enzyme, but much less efficiently than the reversible hydration of CO_2. The esterase reaction, in particular, has been very useful in the kinetic analysis of carbonic anhydrase function, however.

All carbonic anhydrases characterized to date (except the plant enzyme) are compact, nearly spherical, *monomeric proteins with molecular weights around 30,000* and containing *one zinc ion per molecule* (34). A doubling of the carbonic anhydrase gene has probably occurred during evolution, and most mammals, including man, synthesize two distinct enzyme forms, differing in amino acid composition and catalytic activity (Table 2). The nomenclature was originally based on relative electrophoretic mobilities rather than catalytic function or homology, and it should be noted that the bovine form B (Table 2) is structurally and functionally similar to the form C of other species. Cattle seem to lack the low-activity form (36).

Table 2. *Some properties of human and bovine carbonic anhydrases* (34, 64, 66, 85)

	Human B	Human C	Bovine B
Isoionic pH	5.7	7.3	5.65—5.9
N-terminal amino acid	Acetyl-Ala	Acetyl-Ser	Acetyl-Ser
C-terminal amino acid	Phe	Lys	Lys
Cysteine residues/molecule	1	1	0
CO_2-turnover number[a] (k_{cat}, sec^{-1})	2×10^3	0.6×10^6	$\sim 1 \times 10^6$
p-nitrophenyl acetate hydrolysis (k_{cat}/K_m; M^{-1} sec^{-1})[b]	370	2300	970

[a] In 20—25 mM phosphate buffers, pH about 7.5. From data of *Gibbons* and *Edsall* (35) and *Kernohan* (45).

[b] pH 8.9—9.0. Estimated from data of *Verpoorte et al.* (33) and *Thorslund* and *Lindskog* (31).

S. Lindskog

Qualitatively, the physico-chemical properties of the various forms are similar, although they might differ in details. Most of the results discussed in the following sections are thought to apply to all mammalian carbonic anhydrases. It will be indicated in the text where significant differences between forms are known to exist.

1. Metal-Ion Binding and Specificity

Zinc can be removed from carbonic anhydrase on dialysis against a chelating agent at pH about 5 (37, 38). The apoenzyme is inactive but the gross conformation of the protein is maintained (37, 39, 40). The metal-chelating site can accomodate any of the divalent transitional metal ions from Mn^{2+} to Zn^{2+} as well as Cd^{2+} and Hg^{2+} (38, 41). Most of these metallocarbonic anhydrases have low activities or are inactive, however. Only Zn^{2+} and Co^{2+} are efficient activators. As shown in Table 3, this narrow metal-ion specificity is observed for the CO_2 hydration as well as for the esterase reactions.

Table 3. *Metal ion specificity of bovine carbonic anhydrase B*

Metal ion	CO_2 hydration[a]	p-nitrophenyl acetate hydrolysis[a]
Zn^{2+}	100	100
Co^{2+}	50	97
(Ni^{2+}	2	1)
(Fe^{2+}	4	1)
(Mn^{2+}	8	4)

Not active: Hg^{2+}, Cd^{2+}, Pb^{2+}, Fe^{3+}, Be^{2+}, Mg^{2+}, Ca^{2+}, Ba^{2+}.

[a] Relative activities in standard assays (31, 38).

The *stabilities* of these metal complexes follow the *Irving-Williams* series (38), but Co^{2+} and Ni^{2+} are less strongly bound than Zn^{2+} (Table 7). The logarithm of the apparent stability constant for zinc increases linearly and with unit slope between pH 5.5 and 10. No significant differences were found for the human and bovine forms, respectively.

Calorimetric measurements (42) have shown that the *complex-forming reaction*, which involves the dissociation of 1—2 protons (37, 43), is in fact associated with an enthalpy increase of about 8 kcal mole^{-1} at pH 7 (42). The stability is thus linked to an entropy increase of the order 80 cal deg^{-1} mole^{-1} at pH 7.

The *kinetics of zinc binding* has also been studied (*43*). The apparent rate constant is of the order 10^4 M^{-1} sec^{-1} at 25°. The recombination is accelerated by the dissociation of protons from two groups with apparent pK: s of 5.4 and 7.2, respectively. The ionic strength dependence indicates that the metal-binding environment is positively charged at pH 5.7 and 7.8, whereas it is uncharged at pH 8.4. At all conditions the rate constants are several orders of magnitude smaller than for reactions with small ligands (about 10^7 M^{-1} sec^{-1} (*44*)), where the rate of water dissociation from the coordination sphere is rate limiting. Furthermore, the activation energy is 21 kcal $mole^{-1}$ and the activation entropy is about $+28$ cal deg^{-1} $mole^{-1}$ in contrast to reactions with small ligands, where values of 7 to 8 kcal $mole^{-1}$ and -4 to -8 cal deg^{-1} $mole^{-1}$, respectively, are found (*43*).

While these findings undoubtedly reflect properties of the metal-binding site, *structural predictions* based on models are scarcely feasible. The results strongly suggest that the microscopic environment of the active site has a pronounced influence on the reaction. This is, of course, not peculiar to reactions with metal ions but is thought to be of importance for the reactivity and catalytic function of enzymic active sites in general (*45*). However, the large positive values of ΔS and $\Delta S^{\neq}$ suggest that complex formation involves a considerable disordering of water molecules and that the metal ion is partly buried in a groove in the protein. This picture is consistent with the X-ray crystallographic results which will be further described in Sections III A 3 and III A 4.

2. Reporter Properties of Cobalt(II)

The visible absorption spectrum of Co(II) carbonic anhydrase is extremely sensitive to such additions to the enzyme solution which affect the catalytic activity. A striking spectral change occurs as pH is varied around neutrality (*38, 46*). As illustrated for the bovine enzyme in Fig. 4, the isosbestic points suggest the pH-dependent equilibrium of two spectral species, an acidic form with $\varepsilon_{max} = 230$ M^{-1} cm^{-1} at 560 nm showing some structure, and a basic form with four peaks at 520, 550, 618 and 640 nm. Very similar spectra, though with some variations in extinction coefficients, have been recorded for other forms of the enzyme (*38, 40, 41, 47*). The spectral change closely parallels the pH-dependence of the catalytic activity with both CO_2 (*48*) and p-nitrophenyl acetate (*31, 41*) as substrates. The basic form of a group with a pK near 7 is required for activity (Fig. 5). The kinetic behavior of the Co(II) enzyme is qualitatively the same as for the native enzyme (*31, 49*) but the magnitudes of K_m and V_{max} differ (*46*). The pK of the activity-linked group is only slightly affected by metal-substitution, however (*31*). A combination of spectral

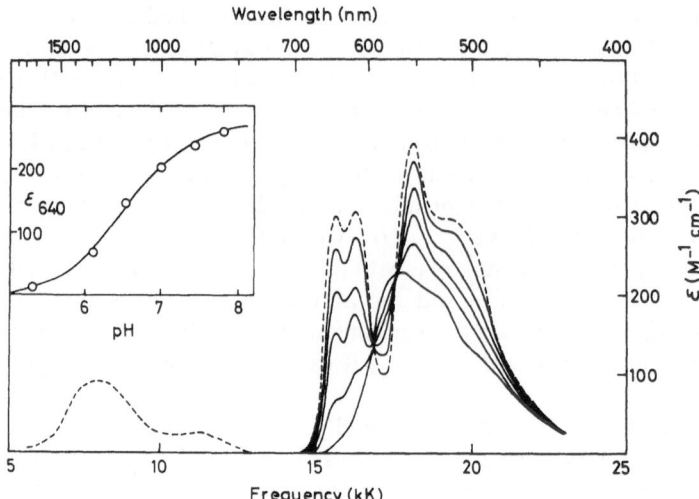

Fig. 4. pH-dependence of the visible absorption spectrum of bovine Co(II) carbonic anhydrase. The broken curve represents the basic spectral form, from ref. (*58*) (near infrared) and at pH 11.6 (visible). The solid curves were obtained in imidazole-sulfate buffers, ionic strength 0.1, pH 7.80, 7.00, 6.55, 6.10, respectively. The spectrum of lowest intensity was obtained by extrapolation and represents the acidic spectral form. Insert: Spectrophotometric titration at 640 nm. The curve was calculated for the titration of a single group with $p\mathrm{K}_a = 6.6$

and kinetic data thus leads to the assignment of the basic spectral species in Fig. 4 as the active enzyme form in the reaction with CO_2 and esters. In the reverse reaction, HCO_3^- is believed to combine with the enzyme form represented by the acidic spectrum of Fig. 4.

Due to the exceptionally fast turnover of the enzymic hydration reaction, it has not yet been possible to study directly the spectral characteristics of any transient enzyme-substrate complexes. The kinetic studies (*48, 49*) show that V_{max} can very considerably without any significant change in K_m. This indicates that K_m approximates the dissociation constant for the enzyme-CO_2 complex. As K_m is practically independent of pH, CO_2 thus seems to bind with about equal strength to both the acidic and the basic forms. The unproductive binding of CO_2 to the acidic form does not change the Co(II) spectrum (*48*). *Riepe* and *Wang* (*50*) found that the infrared absorption of aqueous CO_2 undergoes only a very slight shift on binding to bovine carbonic anhydrase at low pH. They suggested that CO_2 is not bound to the metal ion but is attached to a hydrophobic pocket in the active site.

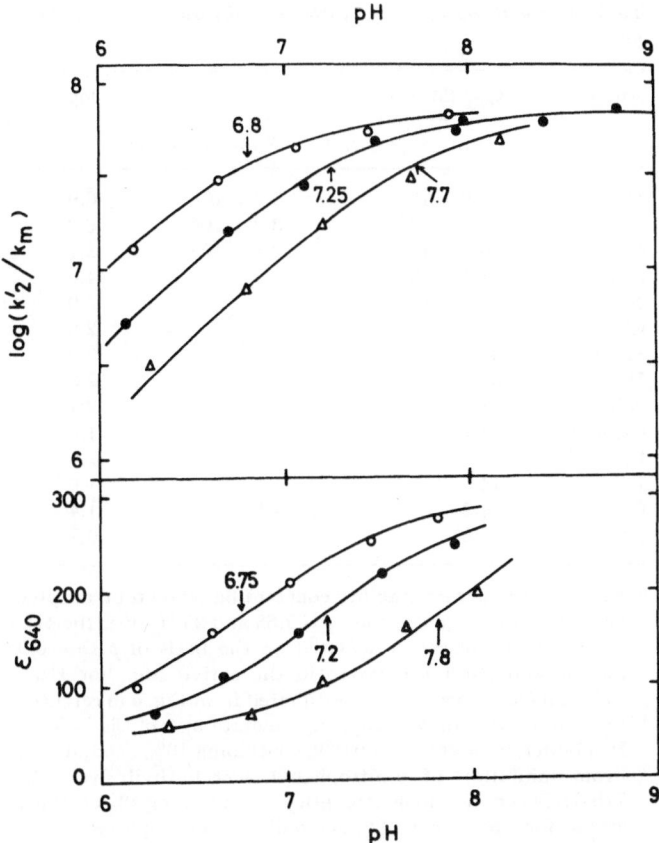

Fig. 5. pH-rate profiles (top) and pH dependence of ε_{640} (bottom) of bovine Co(II) carbonic anhydrase at three Cl⁻ concentrations, from the left 0.025 M, 0.1 M and 0.5 M, respectively. See ref. (*48*) for details

An equilibrium mixture of CO_2 and HCO_3^- at a total concentration sufficient to produce a mixture of enzyme-substrate complexes did not reveal any new spectral species of the Co(II) enzyme (*48, 41*). The spectral change was shifted to higher pH, however, and this was taken to imply a preferential binding of HCO_3^- to the acidic enzyme form with little effect on the cobalt spectrum. Whether this complex represents an active intermediate can not be resolved at present.

Carbonic anhydrase is inhibited by *monovalent anions*. For the bovine enzyme, apparent binding constants range from about 1 M⁻¹ for F⁻ to 5×10^5 M⁻¹ for HS⁻ at 25° C and pH 7.5 (Table 4). The interaction

S. Lindskog

Table 4. *Stabilities of bovine carbonic anhydrase — anion complexes*

Anion	K_{app} (M^{-1}) [a]		log K [a]
	From ref. *(32, 81)* [b]	From ref. *(31)* [c]	
HS⁻	0.9×10^5	5.2×10^5	6.9
CN⁻	3.8×10^5	3.2×10^5	6.9
OCN⁻	0.9×10^4	2×10^4	5.1
SCN⁻	1.7×10^3		4.0
N_3^-	1.7×10^3		4.0
ClO_4^-	62		2.5
HCO_3^-	38		2.3
HSO_3^-	33		2.3
NO_3^-	21		2.1
CH_3COO^-	12		1.8
I⁻	120		2.8
Br⁻	15		1.9
Cl⁻	5.2	4.5	1.4
F⁻	0.85		0,7

[a] K_{app} is the apparent stability constant based on total enzyme and inhibitor concentrations, pH 7.55 and 25 °C. K is the pH-independent constant calculated on the basis of pKa $= 6.9$ for the activity-linked group in the active site. For HS⁻, CN⁻ and OCN⁻, values were estimated from Fig. 4 in ref. *(31)*.

[b] From inhibition of p-nitrophenyl acetate hydrolysis in Tris-HCl buffer, ionic strength 0.009, containing 10% acetonitrile.

[c] From inhibition of p-nitrophenyl acetate hydrolysis. In Tris-H₂SO₄ buffer, ionic strength 0.1, containing 4% acetone, except for CN⁻, where 4% acetonitrile was employed.

of the Co(II) enzyme with these inhibitors results in the formation of new spectral species which are characteristically different for different anions *(46, 48)*, strongly suggesting a close metal-anion linkage. The binding of one anion per enzyme molecule suffices to complete the spectral change (Fig. 6) as well as to abolish the catalytic activity *(48)*. The absorption spectrum of a cobalt enzyme-anion mixture undergoes a spectral change as pH is increased in analogy to that observed in the absence of inhibitor. Like in the case of HCO_3^-, this spectral change is shifted to higher pH values depending on the concentration and binding strength of the inhibitor *(46)*. The basic spectrum of the active enzyme form is always formed at high pH. These results correlate with kinetic data (Fig. 5) showing that the anions bind preferentially to the acidic form, thus shifting the apparent pK of the activity-linked group in the active site.

The most powerful and specific *inhibitors* of carbonic anhydrase are certain *aromatic or heterocyclic sulfonamides* containing an unsubstituted $R—SO_2NH_2$ group (*28*). Inhibitors of this type are of great value in biochemical and physiological studies of carbonic anhydrase action and have found some therapeutic applications (*28*). As the various factors involved in their interaction with the enzyme have been extensively discussed at a recent symposium (*51, 52*) they will only be briefly mentioned here.

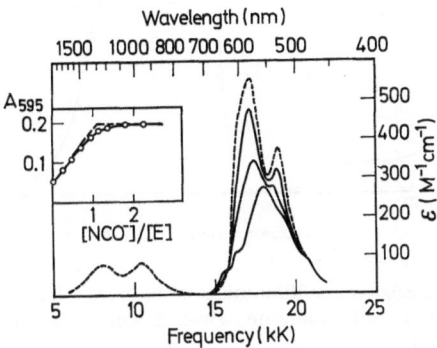

Fig. 6. Spectrophotometric titration of Co(II) carbonic anhydrase with KCNO. Imidazole-sulfate buffer, *p*H 6.1, ionic strength 0.07. Enzyme concentration 0,375 mM. The broken curve represents the 1:1 Co(II) enzyme-NCO⁻ complex. Solid curves: The visible spectra at 0.84, 0.42 and 0 equivalents of NCO⁻, respectively

Although sulfonamides are not known to form stable metal complexes, it has been shown that the presence of zinc or cobalt enhances the strength of sulfonamide binding by several orders of magnitude (*46, 53*). In addition, cobalt enzyme-sulfonamide complexes show visible absorption spectra quite similar to those obtained with certain anions as illustrated in Fig. 7. As further discussed in the following section, a sulfonamide-metal ion contact has been demonstrated by X-ray crystallography. A variety of experimental evidence, such as the *p*H-dependence of inhibitor binding and shifts of ultraviolet absorption spectra and fluorescence emission spectra (*54*), indicate that the sulfonamides are bound through a $R—SO_2NH^−—Me^{2+}$ linkage.

3. Coordination Geometry

Before discussing the wider concepts of the active site and the catalytic mechanism of carbonic anhydrase, it will be considered to what extent the bonding of Co(II) can be predicted on the basis of the spectral evi-

167

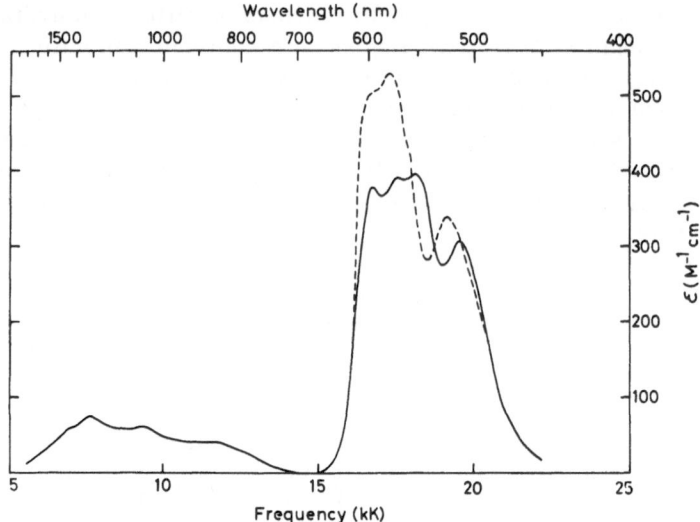

Fig. 7. Absorption spectra of sulfonamide complexes of bovine Co(II) carbonic anhydrase. Solid curve: sulfanilamide. Broken curve: acetazolamide, only the visible spectrum shown

dence. The fact that the spectrum is perturbed when the Co(II) enzyme interacts with known chemical species should make a comparison with model spectra particularly fruitful. *Williams* (*55*) and *Dennard* and *Williams* (*16*) have postulated schemes for the structures of the cobalt-enzyme complexes and proposed a mechanism for the catalysis of CO_2 hydration (see also ref. *56* and *57*). Since the publication of the review by *Dennard* and *Williams*, additional data have been obtained both on the enzyme and on model complexes. These data do not lead to any major revision in their concept of the coordination geometry but some of the conclusions will be brought up here in the light of what is now known about the enzyme structure, and perhaps with a slightly different emphasis.

Measurements of the magnetic susceptibility (*58*) of the cobalt enzyme (Table 5) show that the metal ion is bound as high-spin Co(II). The intensity of the visible absorption makes an octahedral coordination, as well as tetragonal distortions thereof, very unlikely. In the combination with CN^-, the Co(II) enzyme exhibits the spectral features of tetrahedral model complexes with regard to intensity as well as structure both in the visible and the near-infrared wavelength regions (Fig. 8). The width of the near-infrared band (*cf. 20*) indicates that the deviation from tetrahedral symmetry is somewhat greater than in the model shown in Fig. 8 for comparison.

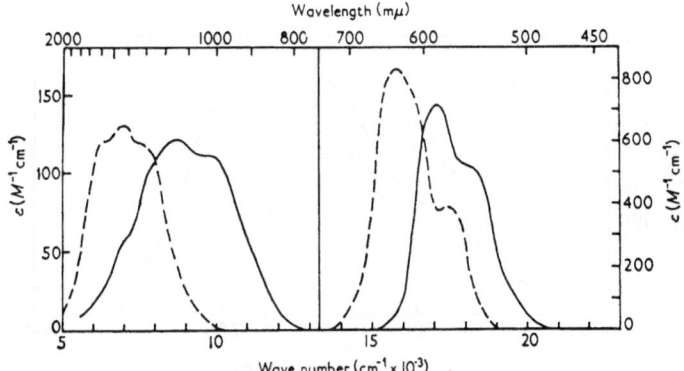

Fig. 8. Absorption spectra of the cyanide complex of bovine Co(II) carbonic anhydrase (solid curve) and the pseudotetrahedral complex, $Co[(C_6H_5)_3PO]_2(SCN)_2$ (broken curve). From *Lindskog* and *Ehrenberg* (*58*) and *Cotton et al.* (*19*), respectively

The complexes with NCO^-, HS^- and sulfonamides all give molar absorbancies from approximately 400 to 600 M^{-1} cm^{-1} which is in the low range of pseudotetrahedral models. Moreover, the bands are broader and show more structure than in the case of CN^- (*cf.* Figures 6—8). These properties suggest substantial contributions from fields of lower symmetry. *Coleman* (*51*) has fitted Gaussian curves to the observed spectra, and his analysis indicates the presence of five closely spaced bands, while the cyanide complex was resolved into three bands.

The spectra obtained with less strongly bound anions (log $K \leqslant 4$, *cf.* Table 4), as well as the acidic spectral form of the free enzyme, have similar band structures (*51*) but are usually even broader and have maximal absorbancies in the range 150 to 250 M^{-1} cm^{-1}. There is no simple correlation between the position of the anion in the spectrochemical series and the absorption spectrum, but the lowest-energy peak of the Cl^--, Br^--, and I^--complexes is shifted towards the red in that order as illustrated in Fig. 9.

The most interesting species from a functional point of view, the free enzyme at alkaline *p*H, shows the most complex spectrum (Fig. 4). The visible absorption is *split into two structured bands* centered at 18.5 and 15.8 kK, respectively. *Coleman's* analysis shows, however, that the

169

major difference in band position between this form and the anion complexes is a shift of the two Gaussian bands of lowest energy (*57*).

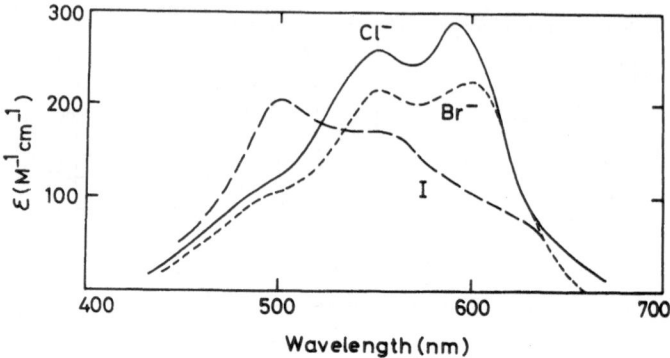

Fig. 9. Absorption spectra of the complexes of bovine Co(II) carbonic anhydrase with Cl⁻, Br⁻ and I⁻, respectively

A similar kind of splitting has been observed in certain supposedly pseudotetrahedral complexes (*19, 59*). *Dennard* and *Williams* (*16*) suggested, however, on the basis of a comparison with spectra of the type shown in Fig. 3, that the Co(II) ion in the active enzyme form is pentacoordinated and that the least strongly colored anion complexes are mixtures of tetrahedral and five-coordinated species. The additional coordination position would in that case be occupied by a water molecule. While the positions of the absorption maxima for the active Co(II) enzyme are rather similar to what is observed in some distorted trigonal bipyramidal models, the correspondence with regard to intensity is less striking. Nevertheless, the spectra do seem to combine features of tetrahedral and pentacoordinated model spectra in varying proportions. This only means, of course, the presence of some specific low-symmetry ligand field, but a firm structural knowledge of the metal-protein linkage is required before the spectral analysis can advance beyond this point. An absorption spectrum practically identical to that of the basic Co(II) enzyme has been obtained in a system containing Co(II), imidazole and phenol (*60*). The structure and composition of the complexed species present in these solutions are not known, however.

The *crystal structure determination* of human carbonic anhydrase C to 5.5 Å resolution (*67*) shows that *the metal ion is* situated near the center of the molecule *at the bottom of a deep crevice* (Fig. 10). Three regions of electron density are found close to the zinc ion. A fourth position can be occupied by the sulfonamide group of an inhibitor (*61*), forming a coordination geometry which may be described as an irregular tetrahedron. Unfortunately, there will probably remain some uncertainty as to possible coordinated water molecules also in the high-resolution (2.0 Å) electron density maps as the region close to the metal ion has been utilized for the binding of heavy atom derivatives and can be expected to be rather blurred.

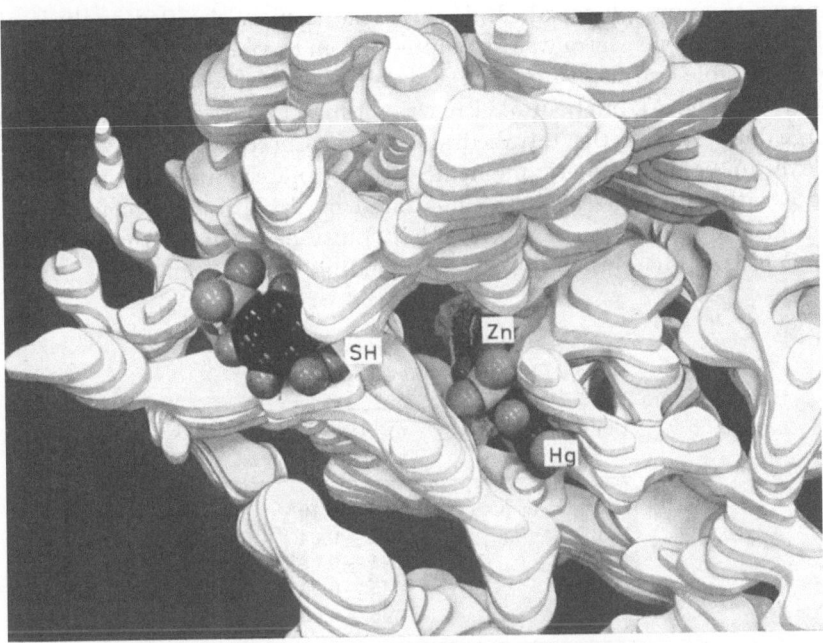

Fig. 10. View of the region around the zinc ion and the SH-group in the complex of human carbonic anhydrase C and acetoxy-mercurisulfanilamide. The model was built on the basis of crystal structure data to 5.5 Å resolution. Two inhibitor molecules are bound; via the Hg atom to the SH-group and via the SO_2NH_2-group to the zinc ion. From *Fridborg et al.* (*61*)

Regardless of what the actual coordination number might be in the active enzyme form, it seems safe to conclude that the *geometry must be quite irregular*. As pointed out by *Freeman* (*11*) and by *Ciampolini* (*13*) for pentacoordinated complexes, minute differences in ligand positions correspond to great differences in formal symmetry, and hence also in the absorption spectra. The variations in geometry indicated by the spectral changes of Co(II) carbonic anhydrase could thus occur without much movement of the ligands provided by the protein, but might principally be due to the number and exact positions of ligands coming from the solution environment of the active site pocket. Possible explanations why the anions may not bind in a uniform manner will be discussed in Section III A 5.

4. The Bonding of Co(II) to the Protein

Since the cyanide- and cyanate-complexes of Co(II) carbonic anhydrase appear to approach tetrahedral symmetry, a more detailed analysis of their absorption spectra was attempted. From the positions of the centers of the visible and near-infrared bands, respectively, a tetrahedral ligand field splitting, Δ_t, of about 5.3 kK was estimated (58). This value is quite large but close to the tetrahedral fields in Co(II)-benzimidazole complexes (62). The magnetic moment of the cyanate complex is in accordance with this value of Δ_t as illustrated in Table 5. To account for a tetrahedral field of such a strength it must be assumed that at least some of the three protein ligands are bonding via nitrogen atoms.

Table 5. *Effective magnetic moments (μ_{eff}) of Co(II) carbonic anhydrase (58)*

Complex	μ_{eff} (B.M., room temperature)
Basic form (OH$^-$)	4.23 ± 0.10[a]
Sulfanilamide	4.72 ± 0.04
NCO$^-$	4.41 ± 0.05[b]
Cl$^-$	4.45 ± 0.08
NO$_3^-$	4.45 ± 0.03

[a] Preliminary low-temperature measurements (*P. Aisen* and *S. Lindskog*, unpublished) indicate a linear susceptibility increase with 1/T between 77 °K and 2.3 °K. Hence, μ_{eff} is practically temperature-independent.

[b] A value of 4.30 B.M. is obtained after correction for a small temperature-independent paramagnetism with the assumption of tetrahedral coordination (12). Compare with [Co(II)-benzimidazole$_4$]$^{2+}$, $\mu_{eff} = 4.28$ B.M. (62) after a similar correction.

Unfortunately, the accumulated physical and chemical data do not yet allow the conclusive positive identification of the ligands, although various candidates have been advanced from time to time (*16, 42, 63*). Amino groups are probably not involved in metal binding, however, as the N-terminal α-amino group is blocked with an acetyl group in the native enzyme (*64*), and the lysyl residues can be chemically modified at the ε-amino functions without effect on the catalytic activity (*65, 66*). The bovine enzyme does not contain any cysteine (*37, 67*), whereas both human isoenzymes have a single SH-group. The spectra of the corresponding Co(II) enzymes are very similar and it was concluded (*38*) that metal-binding must be practically identical in all these forms, excluding sulfur as a ligand also in the human carbonic anhydrases. This was confirmed in the X-ray analysis of the crystal structure of human carbonic anhydrase C. The distance between the SH-group and the zinc ion is about 14 Å (*67*). However, there must be some differences in cobalt coordination for the various forms as the *d-d* bands of the human cobalt enzyme B at high *p*H appears to be optically inactive in contrast to the high-specific activity forms (*39, 51, 68*). The inhibitor complexes are optically active in all forms (Fig. 11). The optical rotatory dispersion and circular dichroism of cobalt carbonic anhydrases have been fully discussed by *Coleman* (*51, 68*).

Strandberg and coworkers (*69*) have collected X-ray diffraction data on the human C enzyme to 2.0 Å resolution, and the corresponding three-dimensional electron density map has been computed. An amino *acid sequence determination* is in progress (*70*) but the interpretation of the crystal structure is at a preliminary stage. The zinc ligands have not been unambiguously identified, but two of them bear the features of imidazole rings (*69*). They occur within the same branch of the peptide chain in a -His-X-His- sequence. The third zinc-protein contact involves a different part of the primary structure but its chemical identity is presently unknown[1]). The Co(II) enzyme has not yet been studied by X-ray methods, but its catalytic capacity is a strong indication that the differences in the binding of the two metal ions are small. Furthermore, the spectral properties of the Co(II) enzyme seem compatible with the coordination geometry observed for Zn(II) in human carbonic anhydrase C.

5. The Anion Binding Site

The X-ray analysis has also been extended to include the iodide complex of human anhydrase C (*69, 71*). This heavy anion was found to be localized at a position very close to the metal ion as had been predicted on the

[1]) Note added in proof: The third ligand has tentatively been identified as yet another imidazole group.

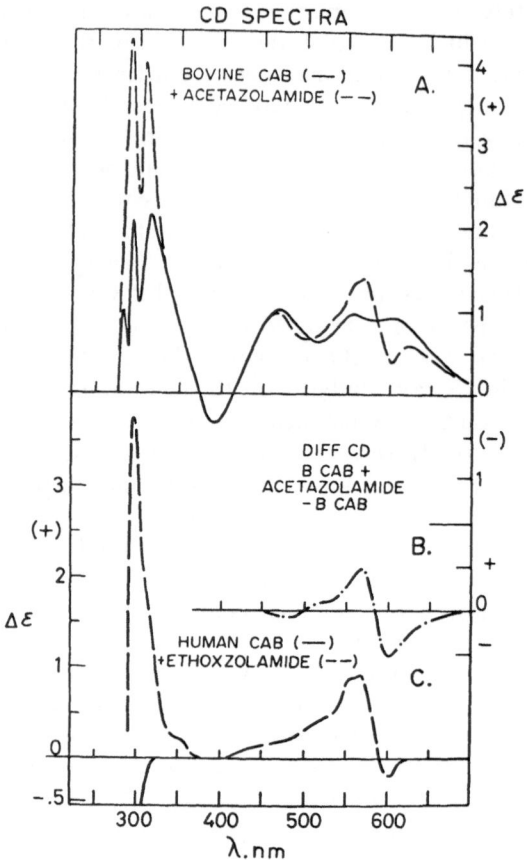

Fig. 11. Circular dichroic spectra of Co(II) carbonic anhydrase and complexes with sulfonamides. A: (————) bovine enzyme B; (----) bovine B enzyme plus acetazolamide. B: bovine B enzyme plus acetazolamide minus bovine B enzyme. C: (————) human B enzyme; (----) human B enzyme plus ethoxzolamide. From *Coleman (51)*

basis of the spectral changes of the Co(II) enzyme. The tentative zinc-iodide distance of about 4 Å appears to be significantly greater than that for small complexes (*ca.* 2.6 Å), however. The anion is pointing from the zinc ion into the deeper part of the active-site crevice. It is not yet known if there is a bridging water molecule between iodide and the metal ion as suggested by *Pocker* and *Dickerson (72)*. These results seem to bear upon the earlier observations that the binding strengths of different anions are too great to be due to a simple metal-ligand interaction in an aqueous medium. The fact that the relative binding strengths of halide ions are in the order $F^- < Cl^- < Br^- < I^-$ (Table 4) in contrast to the reverse order in small zinc complexes (57, 73, 80), also suggest the involvement of additional factors. Moreover, large anions such as NO_3^- and ClO_4^-, coordinating extremely loosely to metal ions in water solution, are tightly bound in the enzyme, whereas divalent anions such as SO_4^{2-} do not inhibit. The observed order of relative effects corresponds roughly to the lyotropic, or *Hofmeister*, series, which is related to the size and hydration of the anions and reflects their effects upon the structure of water around non-polar groups (74).

An anion-binding site with very similar properties has been demonstrated in a non-metal enzyme, acetoacetate decarboxylase from *Clostridium acetobutyricum* (75). *Fridovich* (75) has determined the thermodynamic parameters for anion binding in this enzyme. He found a proportionality between $\triangle H$ and $\triangle S$ of binding. There is an optimal anion size where both these parameters have large negative values. The interpretation offered by Fridovich leads to a picture of the anion site as a *positively charged hole* of a certain size which upon charge neutralization becomes more hydrophobic so that a net ordering of the surrounding water molecules occurs.

The corresponding thermodynamic analysis has not been performed for carbonic anhydrase, but the presence in the active site of another positively charged group has previously been inferred (76, 77). Such anions as CN^- or SH^- forming strong metal bonds of partially covalent character might be bound at a more "normal" distance to the metal ion, which prefers a symmetrical coordination. These anions give intensely colored complexes with the Co(II) enzyme as discussed in Section III A 3. On the other hand, the metal ion seems to interact only weakly with the lyotropic series anions. Their binding must be controlled by the overall electrostatic environment in the anion site, so that the geometry of metal-coordination becomes irregular and sensitive to anion size. Their Co(II)-enzyme complexes are less intensely colored and the spectra are broad and variable (see Section III A 3 and Fig. 9).

6. The Catalytic Mechanism

Until more concrete structural information is obtained, the discussion on the catalytic mechanism of carbonic anhydrase must remain rather speculative. The experimental evidence requires the presence in the active site of a basic group being in some manner linked to the metal ion. This group is generally thought to play a critical role either as a nucleophile in a direct reaction with the substrate, or through general base catlysis. Several schemes for *the function of carbonic anhydrase* have been proposed (16, 41, 50, 78, 79):

175

a) In the simplest model, the pH-rate profile and the corresponding spectral change of the Co(II) enzyme reflect the dissociation around pH 7 of a proton from a metal-bound water molecule (*41, 50*). The reaction of this OH$^-$ with CO_2 leads to a metal-coordinated HCO$_3^-$. Alternatively, the OH$^-$ acting as a general base accepts a proton from a water molecule attacking CO_2. In this case, a water-bridged metal-HCO$_3^-$ complex is formed.

It has been argued (*46, 78*) that these mechanisms are less likely because the pK$_a$ for the process,

$$Me(H_2O)_6^{2+} \;\rightleftharpoons\; Me(H_5O)_5OH^+ + H^+,$$

is rather high, about 9 for Zn^{2+} or Co^{2+} (*80*). However, the hexaquo ions are not representative models for the geometrical arrangement of the metal ion in the enzyme or for the electrostatic microenvironment in the active site. It seems reasonable to assume that the same features of the active site which stabilize an anion-metal ion linkage (see previous section) should also stabilize a metal-bound OH$^-$. However, the basic spectrum of the free Co(II) enzyme is rather unique in comparison with the spectra obtained with other anions, and this has been used as an argument for alternative mechanisms (*16*). On the other hand, OH$^-$ differs from the other anions in its capacity to participate in hydrogen bonding.

Some of the current ideas of the catalytic site of carbonic anhydrase are summarized in Fig. 12. As discussed in Section III A 2, little is known about the binding of CO_2. The possibility of a transient metal-coordinated CO_2 (*16*) is open. No effort has been made to sketch the details of the catalytic mechanism as Fig. 12 undoubtedly will have to be revised after the completion of the crystal structure determination.

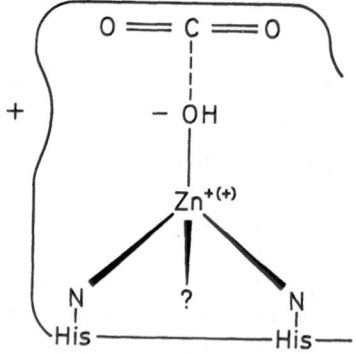

Fig. 12. Tentative scheme of the active site of carbonic anhydrase.

b) Another possibility would be that the metal ion and a proton compete for a protein group in the active site, so that an additional amino acid side chain becomes coordinated at high pH (*16, 46*). To obtain an apparent pK_a of 7 for this process, the side chain must, of course, be much more basic, with a $pK_a \geqslant 11$ (*41, 46, 50*). Thus, histidine can be excluded as shown explicitly by *Riepe* and *Wang* (*50*). On the binding of an anion, this ligand would be displaced and take up a proton. The crystallographic evidence does not favor this alternative, however (*69*).

c) The proton might instead be derived from a protein group further removed from the metal ion but indirectly linked to it, for example, *via* hydrogen bonds to a coordinated water molecule (*78*), so as to control the exact geometry of the zinc or cobalt complex. It is immediately clear that this scheme might differ from the one first described only in the formal localization of charges, and that both represent extremes of one general case.

Pocker and coworkers (*72, 78, 81, 82*) have presented a good deal of argumentation for *histidine* being the activity-linked residue in a mechanism of this kind, and claim that the pK_a for the formation of metal-bound OH⁻ is above 10 in the bovine enzyme. This proposal is not convincing, however, as an imidazole group in a bridged linkage to the metal ion is not likely to have a "normal" pK_a (see paragraph b). Furthermore, a different source for the 1—2 protons dissociated on the binding of zinc to the apoprotein at pH 8 (*37, 43*) must be present. Further studies of the Co(II) enzyme might be helpful in finding alternative explanations for the observed increase in esterase activity at very high pH (*78, 82*). It is already apparent from the spectra shown in Fig. 4 that no drastic change in metal-coordination happens between pH 7.8 and 11.6.

Wang (*79*) has proposed a somewhat different role for an active site histidine residue, involving the promotion of a *proton transfer* from a metal-bound OH⁻ to an oxygen atom of CO_2 (*cf.* Fig. 12) so that a large fraction of the negative charge of the HCO_3^- becomes localized to the O atom in closest contact with the positively charged metal ion. However, in view of the present concept of the binding of this type of anions (see the previous section) such a proton transfer may not be a necessary step in the catalytic reaction.

The presence of *imidazole groups* in the active site region of human carbonic anhydrase B has, in fact, been demonstrated by chemical modification. Thus, bromoacetate reacts specifically with the 3'-N of a histidine residue to give a partially active monocarboxymethyl enzyme (*65*). The reaction depends on the initial combination of the bromoacetate ion with the anion binding site (*65, 83*). In a detailed study, *Bradbury* (*83*) has shown that the irreversible reaction at saturation with iodoacetate

S. Lindskog

requires the basic form of a group in the enzyme-inhibitor complex
having a pK_a about 5.8. This pK_a should represent the reactive histidine
which is thus neither identical with the activity-linked group of pK_a 7
nor likely to be a zinc ligand. The correspondingly modified Co(II)
enzyme has a spectrum of the same type as other anion complexes.
There is a pH-dependent spectral change, but it is shifted to higher pH
compared to the unmodified enzyme. The pH-rate profile of the residual
activity of the modified enzyme is similarly shifted (Fig. 13). These
results are simplest rationalized in terms of the covalent fixation of an
anionic inhibitor.

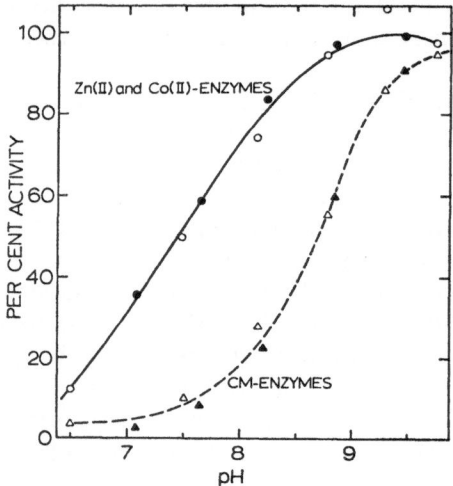

Fig. 13. The pH dependence of the esterase activity of native and carboxymethylated human carbonic anhydrase B. The term CM-enzymes refers to Zn(II)-enzyme
(●, ▲) and Co(II)-enzyme (O, △) which had been carboxymethylated with bromoacetate. Relative activity was estimated in each experiment by taking 100% as the
value corresponding to maximal activity of the same enzyme at high pH. This
value was 17 min^{-1} for the unreacted enzyme and 3.5 and 2.1 min^{-1} for the Zn(II)-
and Co(II)-carboxymethylated enzymes, respectively. From *Whitney et al.* (*65*)

A second histidine residue can be modified at the 3′-position in
human carbonic anhydrase B with a N-chloroacetyl sulfonamide (*84, 85*)
yielding an inactive product.

The mere fact that imidazole is present near the metal ion suggests
that it should have some function, but the data do not allow any unique
conclusion about its possible role in catalysis. Very probably there are

178

structural differences between the active site regions of the human B form and the high specific activity forms (53) as the latter enzymes are not modified by the same reagents. Furthermore, preliminary sequence studies on the human C form suggest that different residues are present in this enzyme at the positions of the modifiable histidines in the sequence of the B form (70). Specific modifications of histidine residues leading to activity losses have been achieved for the human C form with bromopyruvate (86) and for the bovine enzyme with bromoacetazolamide (87), however.

The chemical modification studies have thus not yet led to a much more conclusive picture of the active site than that outlined in Fig. 12, and the identification of *amino acid side chains* involved in catalysis or substrate binding may have to await the completion of the crystal structure determination. The reporter properties of the Co(II) enzyme clearly show, however, that an "open" coordination position is of decisive functional importance, that the metal ion is intimately associated with the basic group participating in the reaction, and that the metal ion is probably also involved in the binding of one of the substrates, HCO_3^-.

B. Carboxypeptidase

Carboxypeptidases A and B are formed by the hydrolytic action of trypsin upon inactive precursors, procarboxypeptidases. These zymogens are synthesized in the pancreas from which they can be isolated (88). Depending on the preparation method, different forms of carboxypeptidase A are produced, varying in the N-terminal region (89).

As indicated by their name, these enzymes are exopeptidases catalyzing the *cleavage of the C-terminal peptide bond* of the substrate molecule. The specificities of the A and B enzymes differ markedly in that the former prefers an aromatic of hydrophobic aliphatic side chain in the C-terminal residue of the substrate, whereas the latter requires a positively charged side chain in the same position (88). Typical low-molecular weight substrates are carbobenzoxyglycyl-L-phenylalanine and hippuryl L-arginine for the A and B enzymes, respectively. The carboxypeptidases also catalyze the hydrolysis of certain esters, such as hippuryl-L-β-phenyllactic acid and hippuryl-L-arginic acid, respectively. Dipeptides having unblocked α-NH_2 groups are hydrolyzed only very slowly.

Bovine carboxypeptidase A has a molecular weight of 34,000 and is composed of *a single peptide chain*. Carboxypeptidase B has similar properties and is probably a homologous protein (90). Both enzymes contain *one zinc ion per molecule*. Carboxypeptidase A has been extensively studied, mainly in the laboratories of *H. Neurath* and *B. L. Vallee*, but the physical properties of the Co(II) enzyme have as yet not been fully utilized. Most of the knowledge of the function of the enzyme rests on other kinds of evidence. However, it falls outside the scope of this review to give a full account of that aspect.

The determination of the crystal structure of carboxypeptidase A and its complex with glycyl-L-tyrosine has been completed to 2.0 Å resolution by *Lipscomb* and coworkes (*91*). A detailed description of the molecule including a comprehensive discussion of structure-function relations have been presented (*91*).

1. Metal Ion Specificity and Substrate Binding

Carboxypeptidase A was the first metalloenzyme where the functional requirement of zinc was clearly demonstrated (*9, 92*). In similarity to carbonic anhydrase, the chelating site can combine with a variety of metal ions (*93*), but the activation specificity is broader. Some metal ions, Pb^{2+}, Cd^{2+} and Hg^{2+}, yield only esterase activity but fail to restore the peptidase activity. Of a variety of cations tested, only Cu^{2+} gives a completely inactive enzyme. In the standard peptidase assay, cobalt carboxypeptidase is the most active metal derivative, while it has about the same esterase activity as the native enzyme ((*93, 94*), Table 6). Kinetically, the Co(II) enzyme shows the same qualitative features as the native enzyme (*95*), and the quantitative differences are not restricted to a single kinetic parameter.

Table 6. *Metal ion specificity of carboxypeptidase A* [a]

Metal ion	Peptidase activity Carbobenzoxyglycyl-L-phenylalanine	Esterase activity Hippuryl-*dl*-β-phenyllactate
Zn^{2+}	7.5	1.15
Co^{2+}	12.0	1.10
Ni^{2+}	8.0	1.00
Mn^{2+}	0,6	0.40
Cu^{2+}	0	0
Hg^{2+}	0	1.34
Cd^{2+}	0	1.75
Pb^{2+}	0	0.60

[a] From *Coleman* and *Vallee* (*93*). See this paper for definition of units and for assay conditions.

The substrate concentration dependence of the cobalt carboxy-peptidase B reactions has also been studied (96). With most of the investigated substrates, the higher activity of the cobalt enzyme as compared to the zinc enzyme is essentially expressed in larger values of V_{max}, while K_m-values are similar.

In the presence of a large excess of Co^{2+}, both native (97) and cobalt (92) carboxypeptidase A show an approximately two-fold activity increase. The kinetics of the enzyme are very complex at moderate or high substrate concentrations and involve both apparent activation and inhibition by substrate (95). Under the standard assay conditions used in connection with the observed cobalt activation, all these complicating factors contribute significantly. The additional Co^{2+} possibly interferes with these secondary effects rather than being a participant in catalysis. Further experimentation is needed to clarify this detail.

Peptide substrates have been shown to bind to the apoenzyme and protect it from reactivation with metal ions (98, 99). The apoenzyme-substrate complexes were estimated to have about the same stabilities as the corresponding complexes with the native enzyme. On the other hand, ester substrates appear to require the presence of the metal ion for binding. Metallocarboxypeptidases, including the inactive Cu(II) enzyme, form complexes with both kinds of substrates hindering the dissociation of the metal ion.

Acetylation and iodination of discrete tyrosyl residues have been achieved, leading to an abolishment of peptidase activity and an enhancement of esterase activity for both native and cobalt carboxypeptidase A (100).

2. Metal Ion Bonding

The 2.0 Å electron density map of carboxypeptidase A shows *three zinc-protein contacts* (91). The ligands have been identified as histidine-69, glutamic acid-72 and histidine-196 (91, 101), where the numbers indicate the positions of the residues in the sequence counted from the N-terminal end. The geometry of the complex is irregular but resembles a distorted tetrahedron with an open position directed towards the active site pocket, and presumably occupied by water in the resting enzyme (91). The similarity with the tentative structure of the metal-binding site in carbonic anhydrase is striking.

While the crystal structure confirms several inferences about the catalytic site based on chemical studies of the enzyme in solution (100, 102), predictions as to metal bonding (16, 103) were less successful. The coordination of zinc (or cobalt) to two protein ligands, an -SH and the terminal α-NH_2, had been postulated principally from studies involving group-specific reagents, determinations of stabilities of metal-carboxy-peptidase complexes, and the absorption spectrum of the cobalt enzyme. It had been pointed out (38, 104), however, that these experiments provided no conclusive demonstration of S,N-coordination. For example, the relative magnitudes of the stability constants given in Table 7 are

Table 7. *Apparent stability constants for metal binding in procarboxy-peptidase A, carboxypeptidase A and human carbonic anhydrase B*

Metal ion	log K_{app}		
	procarboxy-peptidase A[a]	carboxy-peptidase A[a]	human carbonic anhydrase B[b]
Mn^{2+}	3.4	4.6	3.8
Co^{2+}	5.4	7.0	7.2
Ni^{2+}	5.9	8.2	9.5
Cu^{2+}	8.1	10.6	11.6
Zn^{2+}	9.0	10.5	10.5
Cd^{2+}	8.4	10.8	9.2
Hg^{2+}	18.3	21.0	21.5

[a] From *Piras* and *Vallee* (107); pH 8; 4 °C.
[b] From *Lindskog* and *Nyman* (38); pH 5.5; 25 °C.

very similar to the corresponding values for carbonic anhydrase having neither a thiol group nor an α-amino group as metal ligands. Although the absolute magnitudes of the stability constants, K_{app}, compare closely to K_1 for some bidentate ligands, it was not considered that the metal-binding to the protein involves the displacement of a proton at the pH of measurement (93) so that K_{app} must be a function of pH. The results of the measurements of the rates and heat of the reaction between Zn^{2+} and apocarbonic anhydrase (42, 43) discussed in Section III A 1, amply illustrate problems inherent in comparing thermodynamic properties of metal-binding to specific protein sites and to small ligands in water solution.

The visible absorption spectrum of cobalt carboxypeptidase A was at an early stage thought to be characteristic of sulfur coordination (16, 105). The originally published spectrum (94) is poorly resolved (Fig. 14), but a more detailed study has recently been announced (106). Maxima at 555 nm ($\varepsilon = 160$ M^{-1} cm^{-1}) and 572 nm ($\varepsilon = 160$ M^{-1} cm^{-1}), a shoulder at 500 nm and a near-infrared band centered at 940 nm ($\varepsilon = 25$ M^{-1} cm^{-1}) are reported. There seems to be a clear resemblance with spectral forms of cobalt carbonic anhydrase, particularly the chloride complex (Fig. 9). The spectral features seem compatible with the geometry and bonding observed for Zn^{2+} in the crystal structure.

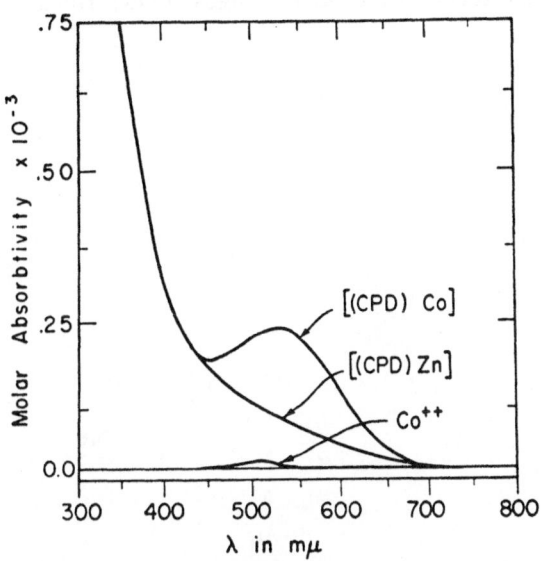

Fig. 14. Absorption spectra of Co(II)-carboxypeptidase A, [(CPD)Co], native carboxypeptidase A, [(CPD)Zn], and Co(H$_2$O)$_6^{2+}$, respectively. From *Coleman* and *Valle* (94)

Metal binding in *procarboxypeptidase A* is weaker than in the active enzyme ((*107*), Table 7). It was proposed that the bonding involves sulfur and a weaker ligand than N (*107*). In view of the present concept of the chelating site in carboxypeptidase, further studies of the zymogen are necessary. In that connection, the cobalt complex should be valuable.

3. The Catalytic Mechanism

The binding of glycyl-L-tyrosine in the active site pocket of carboxypeptidase A is illustrated in Fig. 15. Tyrosine-248 and glutamic acid-270 are believed to participate in the catalytic reaction and represent the acidic and basic groups, respectively, involved in the bell-shaped pH-rate profile. In the bond-cleavage reaction, the carboxyl group of Glu-270 may act by a nucleophilic attack on the carbonyl group while Tyr-248 donates a proton to the amide nitrogen of the susceptible peptide bond. The link between Glu-270 and the α—NH₂ of the dipeptide explains why this substrate is not efficiently hydrolyzed. For a detailed discussion of the structure and possible catalytic mechanisms, the reader is referred to the excellent paper by *Lipscomb et al. (91)*, where also the interesting substrate-induced conformational changes are described.

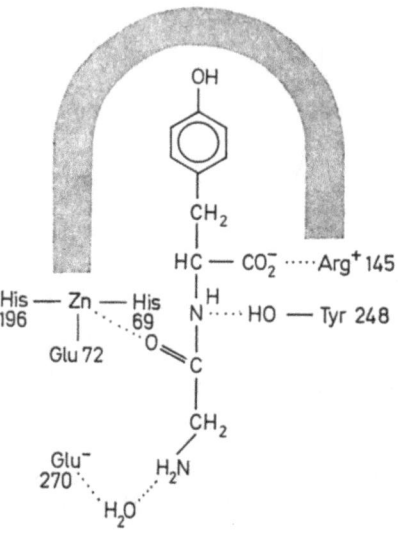

Fig. 15. Schematic drawing of the active site region of carboxypeptidase A interacting with glycyl-L-tyrosine. From *Lipscomb et al. (91)* and sequence information from the laboratory of *H. Neurath (101)*.

The probable *function of the metal ion* includes the binding and polarization of the carbonyl group of the susceptible peptide bond. In the glycyl-L-tyrosine complex, the metal ion is shifted about 0.5 Å towards the substrate relative to its position in the resting enzyme (*91*). Furthermore, *Latt* and *Vallee (106)* have reported that the spectrum of Co(II) carboxypeptidase is changed on the binding of this substrate. This observation should open new possibilities of testing alternative mechanisms on the basis of the crystal structure, and detecting transient intermediates during the catalytic interaction of carboxypeptidase with its rapidly hydrolyzed substrates.

C. Alkaline Phosphatase

Alkaline phosphatases form a widespread group of relatively unspecific enzymes catalyzing the hydrolysis of many orthophosphate monoesters. Their pH optima are generally at pH 8 or above. Several alkaline phosphatases have been shown to contain zinc (3).

The enzyme isolated from the bacterium *Escherichia coli* has been extensively characterized. Its molecular weight is about 80,000 (108, 109). It can be dissociated into half-molecules by acid treatment (108). Reassociation can be brought about at neutral pH. Zinc ions promote this process (108), but dimerization can occur in the absence of metal ions at protein concentrations larger than 1 mg/ml (109, 110). On dialysis of the enzyme against chelating agents a dimeric, inactive apoprotein is produced, showing that the metal ion is also required for activity (111). The binding of Zn^{2+} to the apoenzyme involves some alterations in conformation as indicated by changes in optical rotation and ultraviolet absorption spectrum (110).

The hydrolysis reaction is thought to proceed *via* a covalent enzyme-phosphate intermediate (112):

$$E + S \rightleftharpoons ES \longrightarrow EP_1 + ROH \longrightarrow E + P_1 + ROH$$

where P_1 represents orthophosphate and ROH the alcoholic moiety of the ester, S. The rate of dephosphorylation is linked to the ionization of a group of pK about 7.5 (113) so that the phosphorylation reaction becomes rate-limiting at high pH. At low pH, the intermediate is only slowly hydrolyzed, and a monophosphoryl-enzyme can be isolated, suggesting the presence of one active site per dimer. The phosphate group is esterified to the hydroxyl group of a serine residue (112). However, the possibility of two interdependent active sites should be kept in mind (*cf.* Section III C 2).

1. Metal Ion Binding

The dimeric nature of alkaline phosphatase makes it a more complicated system than carbonic anhydrase or carboxypeptidase. The enzyme contains several metal-binding sites. The stoichiometry of zinc binding is not completely settled. There are at least two strongly bound metal ions (109, 111, 114), but the presence of four specific sites has been claimed (115, 116). At alkaline pH, the enzyme tends to bind even more zinc rather strongly, but probably to sites unrelated to catalytic function (109). A critical evaluation of this aspect falls outside the scope of this review, but it appears that some of the apparent discrepancies are due to different experimental methods in measuring metal binding.

The published values of the *binding strength of Zn^{2+}* show an even wider variation (116—118). Only in one case has an attempt been made to measure the binding directly by equilibrium dialysis with ^{65}Zn and competing chelating agents (117). This study indicated that two strong sites are equivalent and independent. A similar study, where binding was estimated by measuring enzymic activity, suggested a weaker binding for the second zinc ion (118).

The metal ligands are not known, but imidazole has been proposed from titration studies (114) and on the basis of the protective effect of zinc against modification of histidine by photooxidation (119).

In spite of the disagreement in the literature about the number of intrinsic zinc ions in alkaline phosphatase, most data seem to concur that only two sites are linked to the catalytic function, while additional binding may have marginal effects (*cf.*, however, ref. 110).

2. The Cobalt(II) Enzyme

Zinc(II) and Co(II) are the only cations found to reactivate apophosphatase to any appreciable extent (120). The Co(II) enzyme follows the same formal mechanism as the native enzyme, but has a lower specific activity (113, 121). It lacks the phosphotransferase activity (113, 119, 121) observed for the native enzyme, for example in Tris buffers. This was taken to imply that the lower activity of the cobalt enzyme is due to a lower rate of phosphorylation, so that this step becomes rate-limiting also below pH 7 (113). Stopped-flow experiments by *Gottesman et al.* (121) show, however, that a very fast "burst" of p-nitrophenol occurs in the cobalt alkaline phosphatase-catalyzed hydrolysis of p-nitrophenyl phosphate over a wide pH region. These results strongly suggest that a step subsequent to the phosphorylation is rate-limiting in this metal derivative.

Co^{2+} is less strongly bound to alkaline phosphatase than Zn^{2+} (109, 116, 122). The very small stability constant (log $K \approx 4.1$) reported by *Lazdunski et al.* (116) must be considered unrepresentative for the functional ions, however.

Recently, two papers have appeared reporting the spectral properties of cobalt alkaline phosphatase. According to *Simpson* and *Vallee* (115) addition of two equivalents of Co^{2+} to the apoenzyme yields only a small absorption increase around 500 nm, characteristic of octahedral complexes. Only slight activation is observed in this process. Further addition of two more equivalents of Co^{2+} results in full activation and in the formation of an intense visible absorption. Addition in excess of four equivalents has no further effect on activity or spectrum.

Somewhat divergent results, but a practically identical final absorption spectrum, were obtained by *Applebury* and *Coleman* (123), who found that two equivalents of Co^{2+} sufficed to produce a fully developed spectrum. Their spectrophotometric titration is in accordance with two equally strong sites or a pairwise binding of Co^{2+}.

While the question of stoichiometry requires further clarification, there is agreement that the intense spectral form represents *two functionally important cobalt ions* at the active site of the enzyme. Regardless of the existence of other specific sites, the spectral properties of the Co(II) enzyme offer the possibilities of selectively studying the catalytically essential metal-binding sites in this enzyme. Thus, the spectrum depends on pH, and its intensity decreases concurrently with the catalytic

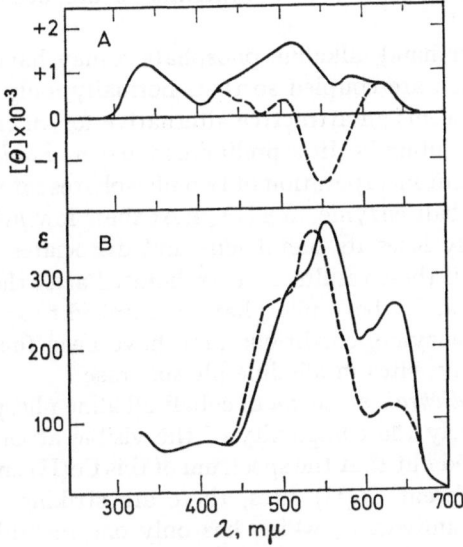

Fig. 16. Circular dichroism and absorption spectra of Co(II) alkaline phosphatase and its complex with HPO_4^{2-}. A: Circular dichroism. B: Absorption spectra. (————), the Co(II)-enzyme. (----), the cobalt enzyme-phosphate complex. From *Applebury* and *Coleman (123)*

activity as the pH is lowered *(115)*. Inorganic orthophosphate, a product of the catalytic reaction, acts as a reversible inhibitor when added to the assay system at high pH, and it can be covalently incorporated at lower values of pH in a reversal of the dephosphorylation step *(115a)*. Phosphate binding at high pH results in a spectral change of the Co(II) enzyme *(115, 123)*, strongly suggesting a close metal-inhibitor linkage (Fig. 16).

Arsenate competes with phosphate and produces a specific spectral change (124).

Whereas *Simpson* and *Vallee* report that *two phosphate ions* are required to produce the spectral change, *Applebury* and *Coleman* find that one is sufficient (*123*) and that two metal ions per dimer are necessary for its binding (*124*). This latter stoichiometry is in accordance with independent observations on phosphate binding to the native enzyme (*110*) as well as the kinetic results indicating the presence of one active site per dimer (*112, 113*). As native preparations containing about three equivalents of Zn^{2+} were found to bind up to 1.4 HPO_4^{2-} per dimer (*124*), some of the observed discrepancies may be due to the different metal contents employed. The extra bound phosphate may be unrelated to the active site, but further experiments are needed to settle this point.

On the other hand, alkaline phosphatase may have two equivalent active sites which are coupled so that, normally, only one can operate at a time. This seems an attractive alternative for an enzyme consisting of two identical subunits. In a preliminary paper, *Lazdunski et al.* (*125*) report the covalent incorporation of two phosphates into the zinc enzyme as well as the cobalt enzyme, at $pH \leqslant 4$. At these low pH values, the free enzyme generally loses its metal ions and dissociates into monomeres (*109*). However, if these results are corroborated after the performance of proper controls, and if both phosphates are linked to specific amino acid residues in the enzyme, conditions may have been found for the "uncoupling" of active sites in alkaline phosphatase.

The *circular dichroic spectrum* of cobalt alkaline phosphatase (Fig. 16) shows more clearly the complexity of the visible absorption. Although it can not be ruled out that the spectrum of this Co(II) enzyme represents two slightly different Co(II) sites, there are striking similarities with Co(II) carbonic anhydrase, which has only one metal-binding site. At high pH, cobalt carbonic anhydrase and cobalt alkaline phosphatase have several spectral features in common, and both may have a similar kind of irregular coordination. It should be noted, however, that the absorption coefficient for Co(II) alkaline phosphatase per equivalent of activity-linked metal ion is only half of the value for Co(II) carbonic anhydrase.

If *E. coli* is grown in the presence of relatively high concentrations (3.5 μM) of $^{60}Co^{2+}$, an active $^{60}Co(II)$ enzyme is synthetized (*122*). Its properties resemble those of the Co(II) enzyme reconstituted from the apoenzyme. Certain mutants of *E. coli* produce defective alkaline phosphatases. The Co(II) enzyme from the mutant U47 lacks the characteristic bands of the Co(II)-complex of the normal enzyme, and the spectrum is broad and poorly resolved (*123*).

D. Alcohol Dehydrogenase

Alcohol dehydrogenases, catalyzing the NAD-dependent oxidation of alcohols to aldehydes, are zinc metalloenzymes (3). The enzyme isolated from horse liver contains two catalytically functional zinc ions per molecule, and two additional zinc ions are believed to have a structural role (126). It has not been possible to prepare a stable apoenzyme, but exchange of $^{65}Zn^{2+}$ for the native Zn^{2+} has been achieved (127). Recently, partial substitution of Co^{2+} for Zn^{2+} in *yeast* alcohol dehydrogenase has been accomplished by growing yeast in the presence of an excess of added Co^{2+} (128). The enzyme isolated from this culture is green, and it is at least as active as the native enzyme (128). This finding should open new possibilities of studying the role of the metal ion in substrate, coenzyme and inhibitor binding.

The absorption spectrum shows three poorly resolved peaks at 620, 670 and 710 nm (Fig. 17). The maximal molar absorbance was about $1000 M^{-1} cm^{-1}$ for the partially substituted preparation which makes $3000 M^{-1} cm^{-1}$ per equivalent of cobalt. The large intensity and the position of the absorption bands at lower frequencies than for the other Co(II) enzymes discussed are striking features of this spectrum, suggesting an irregular coordination of Co(II) in a relatively weak field, presumably involving sulfur ligands. Absorption bands in the near ultraviolet are mentioned (128) which indicates that charge-transfer bands may exist. The available preliminary data to not allow any detailed interpretation, and further spectral and magnetic information would be desirable. An octahedral coordination, which was proposed for the

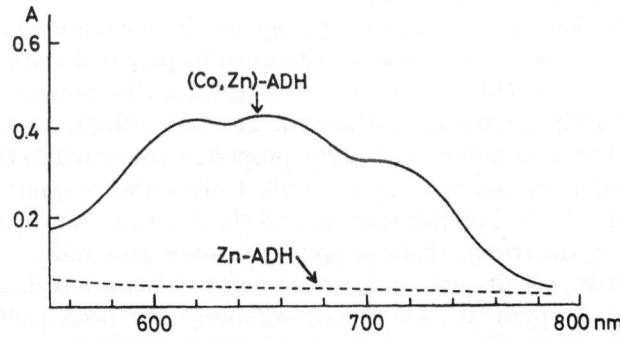

Fig. 17. Visible absorption spectra of partially substituted cobalt yeast alcohol dehydrogenase, (Co, Zn)-ADH, and native yeast alcohol dehydrogenase, (Zn-ADH), respectively. Enzyme concentration, approx. 0.35 mM. From *Curdel* and *Iwatsubo* (128)

catalytically functional zinc ions in the horse liver enzyme (*129*), appears unlikely.

E. Phosphoglucomutase

Phosphoglucomutase, catalyzing the isomerization between glucose-1-phosphate and glucose-6-phosphate via a glucose-1,6-diphosphate intermediate, has very interesting metal-binding properties. The enzyme as normally isolated is partially inhibited by strongly bound heavy metal ion impurities. The apoenzyme is most efficiently activated by Mg^{2+}, and with this weakly bound ion, the enzyme behaves as a metal-activated system. Activity is also obtained with Ni^{2+}, Co^{2+}, Mn^{2+}, Cd^{2+} or Zn^{2+} (*130*). Most of these ions bind quite strongly, so that the enzyme becomes a metalloenzyme according to the usual operational definition. In a series of papers, describing the effects of various metal ions on the activity and ultraviolet absorption spectrum of the enzyme, *Ray* (*130*) and *Ray* and *Peck* (*131, 132*) have also studied Co(II)-phosphoglucomutase. It is mentioned that the cobalt enzyme has a measurable visible absorption (*132*). The spectrum has features reminiscent of Co(II) alkaline phosphatase or Co(II) carbonic anhydrase, and it is sensitive to pH and the addition of substrates and pseudosubstrates (*133*).

F. Other Cobalt Proteins

A number of other cobalt enzymes have been prepared. Their physical properties have not yet been studied, however, and the information is restricted to their catalytic behavior.

The flavoenzyme D-lactate dehydrogenase from yeast has been reported to contain zinc (*134*). An apoenzyme can be prepared and reactivated by Zn^{2+} or Co^{2+} (*135*). When yeast is grown in the presence of added Co^{2+}, a Co(II) enzyme is synthesized. The biosynthetic Co(II) enzyme was found to have different catalytic properties compared to the enzyme reactivated from the apoenzyme (*136*). Only rather fragmentary data have been published on this subject, and the differences in cobalt binding obtained by the two methods of preparation are unknown.

An active, cobalt-containing, oxaloacetate transcarboxylase (methylmalonyl-CoA : pyruvate carboxyltransferase) has been isolated from *Propionobacterium shermanii* grown with $^{60}Co^{2+}$ (*137*). The metal content corresponds to two equivalents of ^{60}Co(II) per mole of enzyme.

Yeast aldolase is a zinc metalloenzyme (*138*). The metal ion is easily dissociated and activity can be restored to the apoenzyme by Zn^{2+},

Co^{2+}, Ni^{2+}, Mn^{2+} and Fe^{2+}, whereas Cu^{2+}, Hg^{2+}, Cd^{2+}, Mg^{2+} and Fe^{3+} fail to reactivate (138).

The zinc ion in a neutral protease from *Bacillus subtilis* has been exchanged with other metal ions (139—141). The Co(II) enzyme is reported to be active (140).

A cobalt complex of transferrin has been prepared by addition of Co(II) citrate to the apoprotein. Hydrogen peroxide was added to obtain the absorption spectrum of cobalt transferrin, and susceptibility measurements showed that the metal ion was incorporated as diamagnetic Co(III) (142).

IV. Concluding Remarks

The most important aspect of the study of Co(II) metalloenzymes is the possibility of using the metal ion as a functional, built-in reporter of the dynamics of the active site. The spectral and magnetic properties of Co(II) carbonic anhydrase have given valuable clues to the catalytic function of this enzyme. The recent studies of Co(II) alkaline phosphatase and Co(II) carboxypeptidase A indicate the general applicability of this approach to enzymes where the probe properties of the constitutive metal ion are poor. The comparison of the absorption spectra of these enzymes and low-molecular weight models have shown that the proteins provide irregular, and in some cases nearly tetrahedral environments. It is obvious, however, that a knowledge of the crystal structures of the enzymes is necessary before the full potential of this method can be exploited.

Metal binding in the cobalt enzymes appears to have certain common features. The tertiary structure of the proteins provides a chelating site where the metal ion is firmly, but not altogether rigidly, fixed. These ligands are not easily exchangeable, and the metal ion is partly "buried" in the protein matrix. At least *one coordination position is open towards the catalytic center*, and can be occupied by a solvent molecule in the resting enzyme or by other compounds which are inhibitory because they interfere with the function of the metal ion. This function involves the binding and polarization of participants in the catalytic reaction, and requires a fast exchange of ligands at the open site.

Possible consequences of the irregular mode of metal binding for the reactivity of the complex have been fully discussed by *Vallee* and *Williams* (17). It is perhaps worth emphasizing, however, that the cooperation with other reactive groups not directly associated with the metal ion seems to be of decisive importance for the catalytic action of the metalloenzymes discussed here.

191

The Co(II) enzymes are still unique in the sense that there are as yet no models of known structure mimicking their spectral properties so as to make possible a direct translation of coordination number, geometry and ligand groups. The spectral properties of Co(II) enzymes suggest that the most useful models from a biochemist's point of view are not necessarily compounds with ligands of the same chemical nature as encountered in the proteins, but compounds of unusual structures and geometries. In fact, there seems to be a trend in coordination chemistry to seek for new and strange ligands forming complexes of different stereochemistries. It is hoped that this review will help to interest coordination chemists in the biochemical aspects of transition metal chemistry and to stimulate them to apply their knowledge of ligand field theory and spectroscopy to ever more complicated complexes, eventually also the metalloenzymes.

V. References

1. *Hogenkamp, H. P. C.:* Ann. Rev. Biochem. 37, 225 (1968).
2. *Hill, J. A., Pratt, J. M., Williams, R. J. P.:* J. Chem. Soc. 5149 (1964).
3. *Vallee, B. L., Coleman, J. E.:* In: Comprehensive Biochemistry 12, 164. Ed. by *M. Florkin* and *E. H. Stotz.* Amsterdam–London–New York: Elsevier 1964.
4. *Smith, E. L.:* In: Mineral Metabolism 2, Part B, 349. Ed. by *C. L. Comar* and *F. Bronner.* New York–London: Academic Press, Inc. 1962.
5. *Malmström, B. G., Rosenberg, A.:* Advan. Enzymol. 21, 131 (1959).
6. *Williams, R. J. P.:* Enzymes 1, 391 (1959).
7. *Vallee, B. L., Riordan, J. F.:* Ann. Rev. Biochem. 38, 733 (1969).
8. *Cohn, M.:* In: Magnetic Resonance in Biological Systems, p. 101. Ed. by *A. Ehrenberg, B. G. Malmström,* and *T. Vänngård.* Oxford: Pergamon Press, Inc. 1967.
9. *Vallee, B. L., Rupley, J. A., Coombs, T. L., Neurath, H.:* J. Am. Chem. Soc. 80, 4750 (1958).
10. *Orgel, L. E.:* An Introduction to Transition-Metal Chemistry: Ligand-Field Theory, p. 71. London: Methuen & Co., Ltd. 1960.
11. *Freeman, H. C.:* Advan. Protein Chem. 22, 258 (1967).
12. *Cotton, F. A., Goodgame, D. M. L., Goodgame, M.:* J. Am. Chem. Soc. 83, 4690. (1961).
13. *Ciampolini, M.:* Struct. Bonding 6, 52 (1969).
14. *Hare, C. R.:* In: Spectroscopy and Structure of Metal Chelate Compounds, p. 73. Ed. by *K. Wakamoto* and *P. J. McCarthy.* New York: John Wiley & Sons, Ltd. 1968.
15. *Carlin, R. L.:* In: Transition Metal Chemistry 1, 1. Ed. by *R. L. Carlin.* New York: Marcel Dekker 1965.
16. *Dennard, A. E., Williams, R. J. P.:* In: Transition Metal Chemistry 2, 116. Ed. by *R. L. Carlin.* New York: Marcel Dekker 1966.
17. *Vallee, B. L., Williams, R. J. P.:* Proc. Natl. Acad. Sci. U.S. 59, 498 (1968).
18. *Ballhausen, C. J., Jørgensen, C. K.:* Acta Chem. Scand. 9, 397 (1955).

19. *Cotton, F. A., Goodgame, D. M. L., Goodgame, M., Sacco, A.:* J. Am. Chem. Soc. *83,* 4157 (1961).
20. *Goodgame, D. M. L., Goodgame, M.:* Inorg. Chem. *4,* 139 (1965).
21. *Ciampolini, M., Nardi, N.:* Inorg. Chem. *6,* 445 (1967).
22. *Bayer, E., Schretzmann, P.:* Struct. Bonding *2,* 181 (1967).
23. *Hunter, S. H., Rodley, G. A.:* In: Proceedings of the XII. International Conference on Coordination Chemistry, p. 4. Ed. by *H. C. Freeman.* Sydney: Science Press 1969.
24. *Cockle, S. A., Hill, H. A. O., Pratt, J. M., Williams, R. J. P.:* Biochim. Biophys. Acta *177,* 686 (1969).
25. *Hamilton, J. A., Blakley, R. L., Looney, F. D., Winfield, M. E.:* Biochim. Biophys. Acta *177,* 374 (1969).
26. *McDonald, C. C., Phillips, W. D.:* Biochem. Biophys. Res. Comun. *35,* 43 (1969).
27. *Keilin, D., Mann, T.:* Biochem. J. *34,* 1163 (1940).
28. *Maren, T. H.:* Physiol. Rev. *47,* 595 (1967).
29. *Pocker, Y., Meany, J. E.:* Biochemistry *4,* 2535 (1965).
30. *Tashian, R. E., Plato, C. C., Shows, T. B., Jr.:* Science *140,* 53 (1963).
31. *Thorslund, A., Lindskog, S.:* European J. Biochem. *3,* 117 (1967).
32. *Pocker, Y., Stone, J. T.:* Biochemistry *6,* 668 (1967).
33. *Verpoorte, J. A., Mehta, S., Edsall, J. T.:* J. Biol. Chem. *242,* 4221 (1967).
34. *Edsall, J. T.:* Harvey Lectures *62,* 191 (1968).
35. *Gibbons, B. H., Edsall, J. T.:* J. Biol. Chem. *239,* 2539 (1964).
36. *Lindskog, S.:* Biochim. Biophys. Acta 39, *218* (1960). — *Carlsson, U., Niklasson, L. G., Svensson, B., Lindskog, S.:* unpublished.
37. — *Malmström, B. G.:* J. Biol. Chem. *237,* 1129 (1962).
38. — *Nyman, P. O.:* Biochim. Biophys. Acta *85,* 462 (1964).
39. — Biochim. Biophys. Acta *122,* 534 (1966).
40. *Coleman, J. E.:* Biochemistry *4,* 2644 (1965).
41. — J. Biol. Chem. *242,* 5212 (1967).
42. *Henkens, R. W., Watt, G. D., Sturtevant, J. M.:* Biochemistry *8,* 1874 (1969).
43. *Henkens, R. W., Sturtevant, J. M.:* J. Am. Chem. Soc. *90,* 2669 (1958).
44. *Eigen, M., Hammes, G. G.:* Advan. Enzymol. *25,* 1 (1963).
45. *Koshland, D. E., Jr., Neet, K. E.:* Ann. Rev. Biochem. *37,* 359 (1968).
46. *Lindskog, S.:* J. Biol. Chem. *238,* 945 (1963).
47. *Duff, T. A., Coleman, J. E.:* Biochemistry *5,* 2009 (1966).
48. *Lindskog, S.:* Biochemistry *5,* 2641 (1966).
49. *Kernohan, J. C.:* Biochim. Biophys. Acta *96,* 304 (1965).
50. *Riepe, M. E., Wang, J. H.:* J. Biol. Chem. *243,* 2779 (1968).
51. *Coleman, J. E.:* In: CO_2: Chemical, Biochemical and Physiological Aspects, p. 141. Ed. by *R. E. Forster, J. T. Edsall, A. B. Otis,* and *F. J. W. Roughton.* Washington, D. C.: N. A. S. A. 1969.
52. *Lindskog, S.:* In: CO_2: Chemical, Biochemical and Physiological Aspects, p. 157. Ed. by *R. E. Forster, J. T. Edsall, A. B. Otis,* and *F. J. W. Roughton.* Washington, D. C.: N. A. S. A. 1969.
53. *Coleman, J. E.:* Nature *214,* 193 (1967).
54. *Chen, R. F., Kernohan, J. C.:* J. Biol. Chem. *242,* 5813 (1967).
55. *Williams, R. J. P.:* Biopolymers Symposia *1,* 515 (1964).
56. — In: Protides of the Biological Fluids. Proceedings of the Fourteenth Colloquium, Bruges 1966, p. 25. Ed. by *H. Peeters.* Amsterdam: Elsevier 1967.
57. *Vallee, B. L., Williams, R. J. P.:* Chem. Brit. *4,* 367 (1968).
58. *Lindskog, S., Ehrenberg, A.:* J. Mol. Biol. *24,* 133 (1967).

59. *Boschi, T., Nicolini, M., Turco, A.:* Coord. Chem. Rev. *1*, 269 (1966).
60. *Dobry-Duclaux, A., May, A.:* Bull. Soc. Chim. Biol. *50*, 2053 (1968).
61. *Fridborg, K., Kannan, K. K., Liljas, A., Lundin, J., Strandberg, B., Strandberg, R., Tilander, B., Wirén, G.:* J. Mol. Biol. *25*, 505 (1967).
62. *Goodgame, M., Cotton, F. A.:* J. Am. Chem. Soc. *84*, 1543 (1962).
63. *Williams, R. J. P.:* In: Proceedings of the Fifth International Congress of Biochemistry, Moscow 1961, Vol. 4, p. 133. Ed. by *P. Desnuelle* and *A. E. Braunstein.* Oxford: Pergamon Press 1963.
64. *Laurent, G., Marriq, C., Garçon, D., Luccioni, F., Derrien, Y.:* Bull. Soc. Chim. Biol. *49*, 1035 (1967).
65. *Whitney, P. L., Nyman, P. O., Malmström, B. G.:* J. Biol. Chem. *242*, 4212 (1967).
66. *Nilsson, A., Lindskog, S.:* European J. Biochem. *2*, 309 (1967).
67. *Nyman, P. O., Lindskog, S.:* Biochim. Biophys. Acta *85*, 141 (1964).
68. *Coleman, J. E.:* Proc. Natl. Acad. Sci. U. S. *59*, 123 (1968).
69. *Strandberg, B., Liljas, A.:* personal communication.
70. *Henderson, L. E.:* In: CO_2: Chemical, Biochemical and Physiological Aspects, p. 121. Ed. by *R. E. Forster, J. T. Edsall, A. B. Otis,* and *F. J. W. Roughton.* Washington, D. C.: N. A. S. A. 1969. — *Henderson, L. E.:* personal communication.
71. *Liljas, A., Kannan, K. K., Bergstén, P. C., Fridborg, K., Järup, L., Lövgren, S., Paradies, H., Strandberg, B., Waara, I.:* In: CO_2: Chemical, Biochemical and Physiological Aspects, p. 89. Ed. by *R. E. Forster, J. T. Edsall, A. B. Otis* and *F. J. W. Roughton.* Washington, D. C.: N. A. S. A., 1969.
72. *Pocker, Y., Dickerson, D. G.:* Biochemistry *7*, 1995 (1968).
73. *Ahrland, S.:* Struct. Bonding *1*, 207 (1966).
74. *von Hippel, P. H., Schleich, T.:* In: Structure and Stability of Biological Macromolecules, p. 417. Ed. by *S. N. Timasheff* and *G. D. Fasman.* New York: Marcel Dekker 1969.
75. *Fridovich, I.:* J. Biol. Chem. *238*, 592 (1963).
76. *Malmström, B. G.:* Federation Proc. *20*, Supplement 10, 60 (1960).
77. *Lindskog, S.:* Studies on the State and Function of Metal Ions in Carbonic Anhydrase. Göteborg: Almquist & Wiksell 1968.
78. *Pocker, Y., Storm, D. R.:* Biochemistry *7*, 1202 (1968).
79. *Wang, J. H.:* In: CO_2: Chemical, Biochemical and Physiological Aspects, p. 101. Ed. by *R. E. Forster, J. T. Edsall, A. B. Otis* and *F. J. W. Roughton.* Washington D. C., N. A. S. A. 1969.
80. *Sillén, L. G., Martell, A. E.:* Stability Constants of Metal-Ion Complexes, 2nd Ed. London: The Biochemical Society 1964.
81. *Pocker, Y., Stone, J. T.:* Biochemistry *7*, 2936 (1968).
82. — — Biochemistry *7*, 4139 (1968).
83. *Bradbury, S. L.:* J. Biol. Chem. *244*, 2002 (1969).
84. *Whitney, P. L., Fölsch, G., Nyman, P. O., Malmström, B. G.:* J. Biol. Chem. *242*, 4206 (1967).
85. *Andersson, B., Nyman, P. O., Strid, L.:* In: CO_2: Biochemical, Chemical and Physiological Aspects. Ed. by *R. E. Forster, J. T. Edsall, A. B. Otis,* and *F. J. W. Roughton.* Washington, D. C.: N. A. S. A., in press.
86. *Göthe, P. O., Nyman, P. O.:* unpublished.
87. *Kandel, S. I., Wong, S.-C. C., Kandel, M., Gornall, A. G.:* J. Biol. Chem. *243* 2437 (1968).
88. *Neurath, H.:* Enzymes *4*, 11 (1960).

89. *Sampath Kumar, K. S. V., Walsh, K. A., Bargetzi, J.-P., Neurath, H.:* Biochemistry *2,* 1475 (1963).
90. *Bradshaw R. A., Neurath, H., Walsh, K. A.:* Proc. Natl. Acad. Sci. U.S. *63,* 406 (1969).
91. *Lipscomb, W. N., Hartsuck, J. A., Reeke, G. N., Quiocho, F. A., Bethge, P. H., Ludwig, M. L., Steitz, T. A., Muirhead, H., Coppola, J. C.:* Brookhaven Symposia in Biology *21,* 24 (1968).
92. *Vallee, B. L., Rupley, J. A., Coombs, T. L., Neurath, H.:* J. Biol. Chem. *235,* 64 (1960).
93. *Coleman, J. E., Vallee, B. L.:* J. Biol. Chem. *236,* 2244 (1961).
94. — — J. Biol. Chem. *235,* 390 (1960).
95. *Davies, R. C., Riordan, J. F., Auld, D. S., Vallee, B. L.:* Biochemistry *7,* 1090 (1968).
96. *Folk, J. E., Clarke Wolff, E., Schirmer, E. W.:* J. Biol. Chem. *237,* 3100 (1962).
97. —, *Gladner, J. A.:* J. Biol. Chem. *235,* 60 (1960).
98. *Coleman, J. E., Vallee, B. L.:* J. Biol. Chem. *237,* 3430 (1962).
99. — — Biochemistry *1,* 1083 (1962).
100. —, *Pulido, P., Vallee, B. L.:* Biochemistry, *5,* 2019 (1962).
101. *Lipscomb, W. N.:* personal communication.
102. *Vallee, B. L.:* Federation Proc. *23,* 8 (1964).
103. *Coombs, T. L., Omote, Y., Vallee, B. L.:* Biochemistry *3,* 653 (1964).
104. *Malmström, B. G., Neilands, J. B.:* Ann. Rev. Biochem. *33,* 331 (1964).
105. *Williams, R. J. P.:* Nature, *188,* 322 (1960).
106. *Latt, S. A., Vallee, B. L.:* Federation Proc. *28,* 534 (1969).
107. *Piras, R., Vallee, B. L.:* Biochemistry *6,* 348 (1967).
108. *Schlesinger, M. J., Barrett, K.:* J. Biol. Chem. *240,* 4284 (1965).
109. *Applebury, M. L., Coleman, J. E.:* J. Biol. Chem. *244,* 308 (1969).
110. *Reynolds, J. A., Schlesinger, M. J.:* Biochemistry *8,* 588 (1969).
111. *Plocke, D. J., Levinthal, C., Vallee, B. L.:* Biochemistry *1,* 373 (1962).
112. *Hummel, J. P., Kalnitzky, G.:* Ann. Rev. Biochem. *33,* 15 (1964).
113. *Lazdunski, C., Lazdunski, M.:* European J. Biochem. *7,* 294 (1969).
114. *Reynolds, J. A., Schlesinger, M. J.:* Biochemistry *7,* 2080 (1968).
115. *Simpson, R. T., Vallee, B. L.:* Biochemistry *7,* 4343 (1968).
115a. *Schwartz, J. H.:* Proc. Natl. Acad. Sci. U.S. *49,* 871 (1963). — *Schwartz, J. H., Crestfield, A. M., Lipmann, F.:* ibid *49,* 722 (1963).
116. *Lazdunski, C., Petitclerc, C., Lazdunski, M.:* European J. Biochem. *8,* 510 (1969)
117. *Csopak, H.:* European J. Biochem. *7,* 186 (1969).
118. *Cohen, S. R., Wilson, I. B.:* Biochemistry *5,* 904 (1966).
119. *Tait, G. H., Vallee, B. L.:* Proc. Natl. Acad. Sci. U.S. *56,* 1247 (1966).
120. *Plocke, D. J., Vallee, B. L.:* Biochemistry *1,* 1039 (1962).
121. *Gottesman, M., Simpson, R. T., Vallee, B. L.:* Biochemistry *8,* 3776 (1969).
122. *Harris, M. I., Coleman, J. E.:* J. Biol. Chem. *243,* 5063 (1968).
123. *Applebury, M. L., Coleman, J. E.:* J. Biol. Chem. *244,* 709 (1969).
124. —, *Johnson, B. P., Coleman, J. E.:* J. Biol. Chem., submitted for publication.
125. *Lazdunski, C., Petitclerc, C., Chappelet, D., Lazdunski, M.:* Biochem. Biophys. Res. Comun. *37,* 744 (1969).
126. *Drum, D. E., Li, T.-K., Vallee, B. L.:* Biochemistry *8,* 3783 (1969).
127. — — Biochemistry *8,* 3792 (1969).
128. *Curdel, A., Iwatsubo, M.:* FEBS Letters *1,* 133 (1968).
129. *Theorell, H., McKinley-McKee, J. S.:* Acta Chem. Scand. *15,* 1834 (1961).
130. *Ray, W. J., Jr.:* J. Biol. Chem. *244,* 3740 (1969).

S. Lindskog

131. *Peck, E. J., Jr., Ray, W. J. .Jr.:* J. Biol. Chem. *244*, 3748 (1969).
132. — — J. Biol. Chem. *244*, 3754 (1969).
133. *Ray, W. J., Jr.:* personal communication.
134. *Iwatsubo, M., Curdel, A.:* Biochem. Biophys. Res. Comun. *6*, 385 (1962).
135. *Curdel, A.:* Comp. Rend. Acad. Sci. *254*, 4092 (1962).
136. — Biochem. Biophys. Res. Comun. *22*, 357 (1966).
137. *Northrop, D., Wood, H. G.:* Federation Proc. *26*, 342 (1967).
138. *Kobes, R. D., Simpson, R. T., Vallee, B. L., Rutter, W. J.:* Biochemistry *8*, 585 (1969).
139. *McConn, J. D., Tsuru, D., Yasunobu, K. T.:* J. Biol. Chem. *239*, 3706 (1964).
140. *Tsuru, D., McConn, J. D., Yasunobu, K. T.:* J. Biol. Chem. *240*, 2415 (1965).
141. *McConn, J. D., Tsuru, D., Yasunobu, K. T.:* Arch. Biochem. Biophys. *120*, 479 (1967).
142. *Aisen, P., Aasa, R., Redfield, A. G.:* J. Biol. Chem. *244*, 4628 (1969).

Received January 23, 1970

Structure and Bonding: Contents Vol. 1-8

Contents Vol. 1-8 (continued)

SPRINGER-VERLAG
BERLIN·HEIDELBERG·NEW YORK

Fortschritte der chemischen Forschung

Topics in Current Chemistry

Herausgeber: A. Davison, Cambridge, MA; M. J. S. Dewar, Austin, TX; K. Hafner, Darmstadt; E. Heilbronner, Basel; U. Hofmann, Heidelberg; K. Niedenzu, Lexington, KY; Kl. Schäfer, Heidelberg; G. Wittig, Heidelberg Schriftleitung: F. Boschke, Heidelberg

Band 15, Heft 1

Orientation and Stereoselection

By Professor **K. Fukui,** Department of Hydrocarbon Chemistry, University of Kyoto, Japan

With 47 figures
85 pages. 1970
Soft cover DM 34,—
US $ 9.40

Die „Fortschritte der chemischen Forschung" erscheinen in rascher Folge in einzeln berechneten Heften, die zu Bänden vereinigt werden. Sie enthalten Fortschrittsberichte über aktuelle Themen aus allen Gebieten der chemischen Wissenschaft. Themen der letzten Hefte waren u. a.: Organische Chemie, Radiochemie, Photochemie, Angewandte physikalische Chemie und Anorganische Chemie. Hauptgesichtspunkt ist eine kritische Sichtung der Literatur (stets zahlreiche Literaturangaben) und Verdeutlichung der Hauptrichtungen des Fortschritts. Auch wenden sich die Fortschrittsberichte nicht ausschließlich an den Spezialisten, sondern an jeden Chemiker, der sich über die Entwicklungen auf den Nachbargebieten zu unterrichten wünscht.

■ **Bitte Prospekt anfordern!**

SPRINGER-VERLAG
BERLIN · HEIDELBERG · NEW YORK

Fortschritte der chemischen Forschung
Topics in Current Chemistry

Orientation and Stereoselection